# Number Savvy

This book is written for the love of numbers. It tells their story, shows how they were invented and used to quantify our world, and explains what quantitative data mean for our lives. It aspires to contribute to overall numeracy through a tour de force presentation of the production, use, and evolution of data.

Understanding our physical world, our economies, and our societies through quantification has been a persistent feature of human evolution. This book starts with a narrative on why and how our ancestors were driven to the invention of number, which is then traced to the eventual arrival at our number system. This is followed by a discussion of how numbers were used for *counting*, how they enabled the *measurement* of physical quantities, and how they led to the *estimation* of man-made and abstract notions in the socio-economic domain. As data don't fall like manna from the sky, a unique feature of this book is that it explains from a teacher's perspective how they're really conceived in our minds, how they're actually produced from individual observations, and how this defines their meaning and interpretation. It discusses the significance of standards, the use of taxonomies, and clarifies a series of misconceptions regarding the making of data. The book then describes the switch to a new research paradigm and its implications, highlights the arrival of microdata, illustrates analytical uses of data, and closes with a look at the future of data and our own role in it.

# Number Savvy

## From the Invention of Numbers to the Future of Data

George Sciadas

CRC Press
Taylor & Francis Group
Boca Raton  London  New York

CRC Press is an imprint of the
Taylor & Francis Group, an **informa** business

A CHAPMAN & HALL BOOK

First edition published 2022
by CRC Press
6000 Broken Sound Parkway NW, Suite 300, Boca Raton, FL 33487-2742

and by CRC Press
4 Park Square, Milton Park, Abingdon, Oxon, OX14 4RN

*CRC Press is an imprint of Taylor & Francis Group, LLC*

© 2023 Taylor & Francis Group, LLC

*Library of Congress Cataloging-in-Publication Data*
Names: Sciadas, George, author.
Title: Number Savvy : from the invention of numbers to the future of data / George Sciadas.
Description: First edition. | Boca Raton : Taylor and Francis, 2023. |
Includes bibliographical references and index. | Summary: "This book is written for the love of numbers. It tells their story, shows how they were invented and used to quantify our world, and explains what quantitative data mean for our lives.
Identifiers: LCCN 2022027085 (print) | LCCN 2022027086 (ebook) |
ISBN 9781032362151 (hardback) | ISBN 9781032357218 (paperback) |
ISBN 9781003330806 (ebook)
Subjects: LCSH: Numeration–History. | Number theory–History.
Classification: LCC QA141.2 .S35 2023 (print) |
LCC QA141.2 (ebook) | DDC 513.509–dc23/eng20220521
LC record available at https://lccn.loc.gov/2022027085
LC ebook record available at https://lccn.loc.gov/2022027086

ISBN: 9781032362151 (hbk)
ISBN: 9781032357218 (pbk)
ISBN: 9781003330806 (ebk)

DOI: 10.1201/9781003330806

Typeset in Palatino
by Newgen Publishing UK

*For my wife Vana,*

*our children Ria and Ioannis,*

*and all the memories at our island*

# Contents

# Acknowledgements

First things first. This book would not have been in your hands without my wife. Quite simply, she's been the heart and soul of the whole endeavour. From the big picture, encouraging me to write this book over the years, to every minute detail that must be dealt with during the transformation of an idea to a final product, her presence was indispensable. Rather than thank you, 'I love you Vana' is more fitting.

Thanks to all those who offered words of encouragement for the writing of this book or help in the early stages of the publishing process. I'm particularly indebted to colleagues with whom I shared various drafts along the way, whether the entire manuscript or parts of it depending on their area of expertise. Their contributions ranged from advice on the organization of ideas to excruciatingly detailed editorial and subject-matter comments which, for the most part, found their way in the final product. The role of George Petrakos, Chris Coward, Fabio Ricciato, Alessandro Alasia, Mahamat Hamit-Haggar, Haaris Jafri, Heidi Ertl, Mark Uhrbach, and Ziad Ghanem is gratefully acknowledged. Thanks also to anonymous reviewers, whose comments too improved the final product. In addition, I'd like to thank my friend Matthew Smith who introduced me to Nola Haddadian. Taking interest in the book's contents, she quickly took charge of the book proposal despite a busy schedule. Her expertise and professionalism proved crucial in the submission process. Getting the help you need at the time you need can never be forgotten.

Thanks are also due to my publishers at CRC. I wish to express my sincere appreciation to my editor, Lara Spieker, who patiently took all our questions and turned them to answers – time and again. Together with Curtis Hill they made it feel that competent help was close by. Thanks also to individuals at Newgen Knowledge Works, particularly the copyeditor, Antony Christ Raja J, and the project manager, Sandrine Pricilla, whose team looked after all kinds of details. Surely, many others contributed to the design of the cover page, the tables and figures, typesetting, or in ways that I may never know – thank you.

Many individuals, if they happen to be among the readers, will unsuspectingly 'see' themselves in this book. They will recall conversations we've had, find answers to questions they asked, or recognize parts of their own thoughts. In that sense, they're all silent partners, as I'm carrying with me numerous and 'scarring' influences from all walks of life, from all continents, and for many years. Whether at the workplace, conferences, or other gatherings, during working hours of after-hours somewhere, I gained enormously from interacting with thousands of immediate and extended team members, colleagues, and students. Those fellow travellers helped shape my views over the years and my love for research and numbers – some more than others. My thanks to all.

# *Author*

George Sciadas has worked in the public, private, and academic sectors. He's well known in statistical circles in Canada and internationally, having worked for more than three decades at Statistics Canada and international organizations, including in several executive capacities. He has also taught at universities for many years. He earned his PhD in Economics at McGill University in Montreal. He has led many national and international projects, with research teams on all continents. He has authored numerous papers and monographs, and has been the editor of influential publications and compendia for many years.

# Author's Log

As late as the closing decades of the 20th century, numerical data were not particularly in vogue. They lived mostly secluded, in research enclaves and policy circles. A few voices advocating for improved numeracy, comparable to literacy long before, were largely falling on deaf ears or frustrated by apathy. I was among those who settled for small inroads, disproportionate to the efforts expended and which frequently made it only as far as a superficial, lip-service mention to data, soon to be relegated to an afterthought in most initiatives. However, partly thanks to those efforts and mostly to what the digital era meant for data, we've come a long ways. The converts to the data cause are now too many. Not surprisingly, this surge to a big tent widened the range of understanding among those involved.

A sure sign of the newfound love affair is that in recent years many are writing books about data. The authors and the readers of these books contribute to our collective numeracy. Some books have a mathematical bent, others continue to focus on the need for more familiarity with orders of magnitude. The height of a stack of crisp dollar bills or the distance defined by putting them next to each other in a row should help differentiate a million from a billion, the capacity of an arena helps mentally visualize a crowd of 20,000 and, perhaps, the number of people standing on each other's shoulders all the way to the moon helps comprehend that distance. While I find much of that useful, this book is not about that. There are plenty of good resources out there. Some books are more technical and aim at methods for professional statisticians and researchers, and some contain findings that promote and encourage more data research. Others yet offer useful advice on the use and misuse of data, and help fight misinformation. I have enjoyed some of those books thoroughly, and some partially. I benefited even from those I didn't quite enjoy, if only because of the thinking they triggered. At numerous times I wished I could either give the authors a pat on the back or talk them out of something – but couldn't step in the books of others. This came with the realization that my intended interventions had less to do with the contents addressed and more with the perspectives of the authors, a few of whom are users of data but many more are observers of the data world – keen observers as it may be. I'm a data practitioner. My affinity with data extends from using to producing them, inside the 'statistical system'. So, here I add my own perspective to the world of data. This book is for everyone with an interest in quantitative information, including mature and aspiring researchers and their audiences.

## Numbers and I

In full transparency, I do have a very long and deep relationship with numbers. A relationship of reciprocal attraction and mutual respect. I'm not an objective observer looking unemotionally from the outside in, and freely confess to – and own – whatever biases this may entail. My lifetime rapport with data has created a strong bond and an empathetic understanding between us. Data talk to me often, and mostly I listen. Other times I talk to them, using all range of emotion. We express ourselves freely, with good manners. We can take each other's bad jokes. We use, but never misuse, each other. Data use me to promote themselves and showcase their prowess. I use them to understand issues and, here and there, to have fun. Once in a while data praise me for how I treat them (flatterers) but at times they may grudge, moan, or grin. We'll never *ever* boss each other around and we always have each other's back. When occasionally we're apart, we miss each other terribly.

Numbers have given me so much. Most times, despite the volume of information I may have in front of me, my level of understanding of an issue refuses to budge, to evolve. Until I see numbers, that is. Then, somehow, I click. Everything falls into place. I can even explain to others! With numbers I understand the true essence of things.

**George Sciadas**
*Summer 2021*

# Welcome

Welcome to our meeting point for your visit to Arithmoland, the land of numbers. My name is George and I'll be your guide. Some of you had long flights, others came by land. Now we're only a short drive away from our destination, the land where numberpeople work and live, surrounded by tourists. The place is huge and constantly expanding, there's no end in sight. I've seen many areas myself, even lived there for a period, but can't quite tell you. Sure thing is that the place is booming, it's bustling with activity, day and night. Everybody now cares about data, you see...Legend has it that some Arithmolanders never sleep, but that's a myth. Hype, you know...

Our itinerary choices reflect my own interests and knowledge of the place. There's always more to see and to do. We'll drive by some areas for a quick look and we'll explore others deeper. We'll visit upscale neighbourhoods with wide avenues and well-manicured data gardens but also downtrodden and neglected dark alleys with foul odours emanating from data dumpsters. We'll browse at old and new architectural landmarks, admire famous data skyscrapers, even stop for some data ingestion to experience everyday life. We'll enter galleries with spectacular visual shrines to winning numbers, throw tokens in data fountains and make a wish, look at statues of data heroes long departed, even lay a wreath in the *Tomb of the Unknown Number*. We'll see locals hard at work, producing and using data, many trying to find their way through the congestion of data traffic in the downtown core, others sipping coffee in more leisurely data suburbs. We'll then venture all the way to the periphery and take a peak at how things are unfolding and what may be in store for the future. You'll find that most folks are very friendly, knowledgeable, and eager to engage in conversations about data. A few tend to be more reserved. All those tourists with the cameras, I guess...but you can take as many selfies as you like.

One more thing. The nearest onramp to the highway is closed, *big data* construction project, you see – what's new. So, we'll take a brief detour. I know detours can be annoying. Narrow and winding roads, can't go straight at all, can't see far in front of you, you feel you're not going anywhere. But, if it's any consolation, we'll enter from the ancient area where barely anyone goes anymore, the birthplace of number. Then we'll connect further out on the highway and we'll practically be there. Along the way I'll share a few things that we normally don't think of, in a light-hearted manner. Most will not raise eyebrows but some may surprise you here and there. They will all come handy later, though. I hope you enjoy the ride.

# 1

## We Got Number

Since this is a book about numbers, I'll start from the beginning. Where did numbers come from?

### Where It All Starts

Our quest for answers is the story of human civilization. The invention, development, and use of numbers are part and parcel of human evolution. Numbers are indispensable to scientific and socio-economic understanding, and they represent a fantastic tool in the search for answers.

We first experience our world through our senses. Our vision enables our innate ability to distinguish certain qualities around us. We can tell shapes, colour, and texture – although not with the stereotypes we now have in mind. And, yes, we can tell *quantity*. More like another quality, though. Our eyes alone can tell that there's a quantity difference between a bunch of trees and a dense forest, a very wide river and a crossing to the other bank you can simply jump over, a mountain peak high in the sky, and a little bump on the ground that we should be careful not to trip over. With enough observations, we can sense big and small, high and low, closer and farther – up to a point. Such sense of quantities communicated through our senses is not quantification. We're not born with an organ that can tell numbers or really quantify the world around us. In fact, our innate ability for numeracy is miniscule, practically non-existent. Surely, we can discern very small quantities – and so do other animals – but this doesn't extend to larger numbers. If we're shown a card with nine trees, we're incapable of making out the number without counting. Through plenty of experiments over time among infants, adults, and primitive tribes, experts tell us that we can be reasonably expected to tell quantity up to four. Not a stellar performance, particularly if we want to build organized civilizations and have ambitions to go to the stars. So, if not in our physiology, where does it all start? How come we have numbers? How did we end up with integers, fractions, percentages, ratios, indexes, big data, and the like? What is *the thing* between quantity and number? Well, it's an interesting story! Let's have a look.

DOI: 10.1201/9781003330806-1

The relationship of humans with numbers has been a long and arduous one. It's a safe bet that the first notion that sank in was the singularity, the One. If nothing else, it must have been brought about by self-awareness – me, as distinct from the others and the surroundings. Even then, *oneness* would not have yet been perceived at an abstract and generalizable level in a way that it can be applied regardless of context. In other words, it was not a number. For instance, seeing only one of something the early human or the new toddler would experience person, tree, river, dog but not one person, one tree, one river, one dog. Either way, there's only so much you can do with the *one* alone.

Our vision may be an imperfect measuring instrument but, in addition to ballparking quantities, it's an amazing data gathering device. Much like today, this is done through *observations*. So, early humans continued to observe. Our bodies offered many useful data points for small numbers. From eyes and ears to limbs, fingers, and toes. Nature offered many more examples, both of small and large 'numbers'. The cloverleaf, the legs of the animals, a few stones lumped together, and a whole big bunch of trees (forest). The sun and the moon stand out amongst the stars that have no end in sight, as do a few boulders on the beach against the grains of sand. Our senses can tell apart singularity from endlessness and make us aware of the range that exists in between.

Still, to move to *two* more abstraction is needed. Perhaps, it started with duality and the recognition of *a pair*. Our eyes, ears, and hands appear identical. The two robins you see everyday look very much alike and always together. But, even if side by side, a little tree with a smooth bark and light green leaves isn't alike a large conifer tree with a rough bark and dark green needles. You can't quite think of them as *two*, let alone group them in pairs conceptually similar to those of two eyes. It's hard to think of an apple and an orange as *two* fruits (we still refer to 'apples and oranges'). However, the more we observe the more we discern *likeness* as much as differences – at least in a taxonomical sense. The two trees are more alike than any of them is with a nearby boulder or a flying eagle. So, repeated exposure to observations, combined with likeness, could have led to the notion of a pair, even to two. We'll never quite know but it seems that such a process didn't lead to three, four, and beyond. There are numerous accounts by anthropologists and ethnographers who, for more than a century now, tout the *one-two-many* cultures they encountered and studied in tribes from the Amazon to Africa and the Pacific. Their languages contain words for one and two, moving straight to *many* after. That's the extent of their numeracy.

Sensing quantity, in terms of differentiating orders of magnitude, exists even in animals. A pack of wolves know they outnumber a single gazelle and dare to attack, something unthinkable on a one-on-one situation. Being able to ascertain quantity at an approximate level, such as *more* or *less*, was perhaps adequate when human tribes were hunters on the go. Why care about exact quantities? A pack of wolves at a distance would elicit a behaviour, such as taking precautions. But it would matter little if the pack was 15 or 18-strong. It may be a truism to say that evolution responds to *needs*. The inescapable reality is that an invention must satisfy some need. A wheel to move around easier, a fishing rod to catch fish. Exact quantification enabled by numbers also needs to be useful. What need will the invention of number fill? We'll get to that next.

The earliest archaeological finding related to numbers is dated at about 35,000 BC, and it's a fossil of an animal bone marked with notches. Many more similar fossils from later periods have been found in Africa, across Europe, and elsewhere. These fossils coincide with the times when humans started to stay put and form small-scale agrarian societies and are believed to represent flocks of animals herded by shepherds. Piles of pebbles and other objects that seemed to have filled the same purpose have also been found. It's understandable that someone cared to have a better account of their sheep than for a distant pack of threatening wolves. Other than those discoveries, we had to wait until the beginning of history (around 3,000 BC) to find lots of artefacts that contain evidence not only of numbers, as more or less we know them today, but also of counting, record-keeping, and calculating. From the Sumerians in Elam, and later from India, China, Egypt, Greece, and Rome, it's clear that *number* has arrived. So, numbers were invented somewhere in the period between the early notched bones and the first existing historical records – and I suspect much closer to the beginning of that period.

Researchers believe that it's highly unlikely that those notched bones represent the first numbers. Such tallies were early efforts to find out what was missing. A rudimentary inventory management! Same was the function of piles of pebbles (*calculi* in Latin) or other found objects that were also used to account for flocks of animals, perhaps tools, weapons, or something else. It's true that to accomplish such a task we don't really need to know what number is or how to count. All we need is to trace the evolution of a given set over two points in time. A notch or a pebble for every animal at the beginning of the day, marked against another at the end of the day. Doesn't matter if we started with 35 and ended up with 33 or with 25 and ended up with 23. Like musical chairs in birthday parties, you know a kid will be left standing without a chair. How many exactly are sitting doesn't quite matter. The game can go on. True number is an abstraction, not a tally.

## An Ancestor's Epiphany

I can follow the reasoning above but I'm not satisfied that someone can remain trapped at those tallies, without number, for long. That's why I feel that the period from tallies to real number was relatively short, not tens of thousands of years. Several reasons come to mind.

To begin with, it's quite possible for a hunter to use a bone and carve notches that account for his kills claiming bragging rights. He'll do this at different times and even carve different sets of notches for different types of animals (some archaeological finds seem to corroborate exactly that). A big boar gets one notch, a small bird gets a smaller one below, like ledger entries in series of ones. The shepherd will carve all notches at the same time, presumably in the morning when the sheep passed through the gate on their way out to graze. Having already abstracted the notion of one, a big ewe gets one notch and so does her little kid. But what exactly did he do at the end of the day? Did he cross the notches he carved in the morning to see if any was left uncrossed? Did he carve another notch underneath each one? Either way, his interest was surely on the difference between the two, that smaller 'number' of missing animals. How many notches short was he? When he went out looking for them he'd know if he had to fetch one, two, or more sheep. He would identify the difference, the excess notches, as number. The outcome would be similar when comparing his notches with those of other shepherds in the tribe, something bound to happen sooner or later. The difference in the series of notches would reveal a quantity.

Then, there's the question of the next day. Will he start all over? And for how long, especially if he mastered the difference between tallies? More importantly, how did he deal with flock attrition and births? Scratch (erase) some notches and add others, net them out, or start all over on a new bone after each event? The point is that soon after the establishment of the stock itself, it becomes necessary to develop practical approaches for routine monitoring, including additions and subtractions. Anything else is too cumbersome and unproductive – and there's no evidence of bones with multiple scratches on top of scratches and the like. For how long can he miss the notion of an actual herd size, especially having mastered series of ones and their differences? Hard to stay at that stage for tens of thousands of years. Indeed, tallies started to be expressed differently, such as scratching four notches for the five-bar gate (卌). If not a sure giveaway for actual counting rather than tallying, such shortcuts are eminently conducive to counting. You may relate to that from the set theory in your algebra courses. One-to-one associations are quite useful up to the point you need to know a 'number', the quantities of the units contained in the sets and their difference – as distinct from which set is larger. (It may be that our brains today are conditioned differently, but how many times you'll bring out enough chairs for all-minus-one kids without counting how many chairs you need before you too re-invent *number*?)

There's an additional reason behind my belief that the invention of number happened earlier than later. If the number had not been invented, as languages were evolving they would have to develop different words for two horses, three horses, and three cows. While many languages have words dedicated to any number of detailed things, including winds based on their direction and different types of waves, they don't contain words for even small sets of different animals or things. A word and a number are sufficient to express all we want.

Our ancestor's observations, groupings based on likeness, familiarity with the oneness and its repetitive depiction that led to the creation and manipulation of sets, left many traces. Regardless of the exact timing, the sure thing is that at some point our ancestor was hit by an epiphany. He connected all those traces and found *number* as an abstraction beyond the series of notches. This new entity emerged in his head. Notches are just a visual aid, he didn't really need them or the pebbles. He jumped up with a high leap boosted by his eureka moment. I found number! I can count! I can add one to the one and get two, one more for three, and I can go on for any number I like. That's the time when the abstract construct of number takes over from the perceptual collection of objects. Now we can move from general plurality to precise quantities, from more or less to exact counts, from counting sheep to counting anything. The abstraction is complete. (The *unit* is not needed yet, later it will become subservient to the number.)

The bonfire party that surely happened that night was not unique, though. Our ancestors with the epiphany were many, at different places and different times. Still, now he has power. He can think of numbers that he never knew because they don't exist around him, they're in his head. His imagination goes wild. What if he could take matters further? Do what central commands and even free markets try to do. Use births and deaths to calculate the stock of sheep over time, put it against estimates of annual tribe consumption factoring in projected tribe size, and arrive at a sustainable, long-term, optimal herd stock. He'd tell the others in the tribe, other tribes too. Not only he'd be *the man* at the bonfire parties but that would surely guarantee him glory for life! But such enthusiasm would have to come down a notch – for a while.

## From Number to Counting and Measuring

So, we got number. Big 'moment' in our history, a major breakthrough! But where does this lead? What do we do with such a new and powerful invention? Well, what people always do. We put the invention to use, good and evil. For sure the new invention liberated creativity, people started to count. They counted what they could, animals, people, belongings. As they became good at counting, they invented many *ones*, different units to measure distance, weight, and volume, extending the natural ability of the eye. They even realized that the exact unit doesn't quite matter, it's another abstraction. Any *one* will do. Even the day became a one, a unit in the lunar cycle. Imagine, assigning a number to a natural phenomenon with fixed periodicity, measuring and keeping track of time!

Besides enabling counting and measurement, numbers prompted basic calculations. By the time of the Sumerian historical records there had certainly been many, simple accounting too. Naturally, they were limited to quantities that made practical sense in the context of those societies. Following the advancement of the economies of the time, they extended to transactions involving exchanges of goods and payments, credits and debits, assets and liabilities. The needs were generally manageable, and there was no need for very large numbers until people got interested in astronomy. (Even today we struggle with large numbers.)

Even with counting, measurements, and simple calculations, we were still far away from our destination – a functioning number system. One that would truly free the awesome power of number. More, much more, was to come. Thankfully, from then on, there are enough archaeological discoveries from long-gone civilizations that our story can be less speculative. From notched bones, to sticks and pebbles, to bullae of clay tokens, inscriptions on stone and papyrus, records started to be kept. And they contain numbers – and then some.

## A Triple Problem

Back to our ancestor. He invented something big, surely, now he knows. But what about the others? To put the newfound numbers to good use, others would need to know too. He has to tell them all about number and teach them. In the process, he needs to create words for the numbers, give them names – symbols too. Notation is key for record-keeping, and later it'll help carry out calculations that can't be done mentally. It's possible that our ancestor started to write down numbers even before letters.

If you think these were big-enough problems for the day, just wait. In the process of solving them, he'll inadvertently have to deal with a closely related but thornier issue. One that he's not even fully aware of but which lurks in the background and gets in the way. The choice of a *base* for a number system. Rather than taking on these problems one at a time, let's see how naming, notation, and base were dealt with in parallel. How did our ancestor engage in educational and communication campaigns while still working hard on further developments? Having come thus far, clearly our ancestor is a thinker. So, he thinks…as if he were you.

How would *you* start naming numbers? The one-two-many tribes made up words for one and two and then combined them to express slightly higher numbers, like two-one for three, two-twos for four. You try it too but very soon you realize its limitations. Imagine having to say two-two-two-two-twos and only be at ten. You're better than that. You start crafting names for the three, the four, and the five. How far do you go? To 10? What about 23? Do you dare 97? And what do you do after? How many words can you and your tribe really handle and remember? And, from all you've seen so far, this number thing doesn't seem to end, there's always another one coming. You may wonder if you invented a monster...

As you ponder the name conventions, you experience similar issues trying to jot numbers down. It feels intuitive to draw a stick (vertical line) for one. It works well – and it's the same we still use. The notch, the stick, the numeral 1, or anything else is a mere dress for the real abstract number underneath. As with tallies, you continue with two sticks for the number two, three sticks for the number three, and so on. Most known early systems resorted to lines and dots. Same as children draw stick people, nothing more imaginative. Proof that numbers were developed from scratch by humans, no extraterrestrial intervention here! Notation symbols were used from right to left, left to right, even from up to down. They all had several versions over time. Still, they were economical in their choices. The Sumerians and Babylonians put the sticks in cuneiform and only used symbols for one (𒐕) and ten (𒌋), the early Greeks slanted them somewhat. The Chinese (mostly) had the first three lines horizontal and used dots too. Later, the Mayans used dots and bars (like the Morse code) but also elaborate artistic symbols for large numbers given their interest in astronomy. The Egyptians used pictorials (of course), and different ones for different occasions. As a testament to the significance of these efforts, Figure 1.1 shows one of the images encoded in the *Golden Record*, a time capsule sent permanently to space with Voyageur

**FIGURE 1.1**

Communicating with extraterrestrials

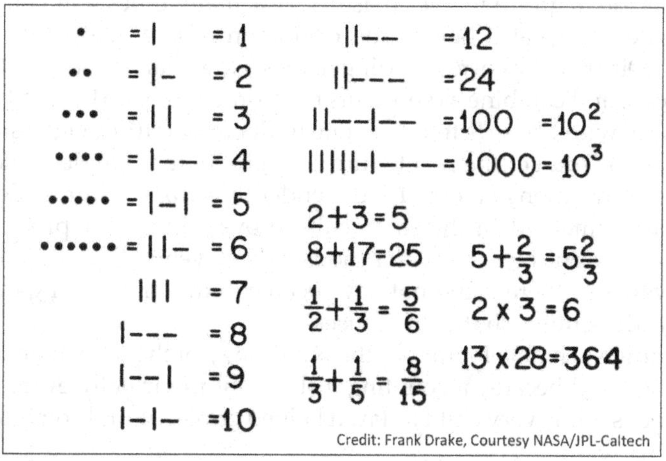

Credit: Frank Drake, Courtesy NASA/JPL-Caltech

in 1977.[1] This 'bottle in the cosmic ocean,' as Carl Sagan called it, is intended to introduce human life on earth to any extraterrestrial intelligence it may encounter out there.

For the most part, symbols were graphical representations of the quantities they represented. (We still recognize them today in many Roman numerals.) Still, very soon you grunt. How far do you carry this? Examples up to nine abound but it's a matter of time before you realize that it's unworkable, especially since you can't quite make out the number of sticks after four. Even with a seven you'll have to count the sticks every time you want to know the number. It sort of defeats the purpose. But you don't give up. You try using your fingers…one for the left thumb, two for the index, five for the pinky. You get the clue to group in fives, perhaps tens if you use both hands (rudimentary bases). Practically, rather than adding a stick after four, you may cross them for a five (卌). Not only this five-bar gate works well but it became famous for millennia and was still used not too long ago. Then, you add sticks again for the next numbers and can choose to call them five-one and five-two or six and seven. When it comes to ten, you can add a second cross to the five-bar gate, use two of them, or another symbol altogether – a circle, or an X. Similar is the issue of names. Regardless of how you notate the ten, you can still call it two-fives or ten. If you decide to group in tens, ten can have its very own name, which can be then used to name the 11 (ten-one) and the 12 (ten-two).

This interplay between naming, notation, and base is neither a trivial nor a straightforward process with a unique solution. There's no prescribed way to go about it and you can appreciate that for the ancients the task was daunting. All kinds of decisions have to be made along the way, each with its own pros and cons. The possibilities start to feel endless. So, you continue experimenting to the chagrin of those around you who start to see you as eccentric. ("*He's spending way too much time on that stone desk of his inside his man-cave*".)

If you decide to group in fives (base 5), how many times will you cross the five-bar gate or repeat it before you decide on new symbols for its higher multiples? Ditto for the corresponding names. How many words and symbols do you create and combine as shortcuts for more? Today's 51 could have been ten-fives-one with a base 5, five-tens-one under a base 10, or eight-sixes-three using a base 6. You get the point. Sure enough, there is ample evidence that numerous conventions emerged independently across cultures. Competing versions also emerged in the same culture over time. This process wasn't short-lived, it went on for several thousands of years. Given the historical records available, there could not have been coordination among them. Still, the similarities outweigh the differences.

Throughout such developments, the physiology of the human body proved crucial. The hand became a counting instrument in virtually every culture – and continues with every child today. It helped cross-cultural communication.

It was used with different degrees of sophistication and was frequently combined with other body parts. Using the knuckles, toes, knees, and elbows, some cultures could count up to 1,000. Through an ingenious system, the Chinese could count up to 100,000 in one hand and up to a million with both! Many learned to use the hand and carry out elaborate calculations too. (In *The Universal History of Numbers*, Georges Ifrah offers a fascinating account.[2])

With the arrival of alphabets things took another turn. Some introduced notation based on letters, notably the Greeks and the Hebrews. Greeks, who had always favoured a decimal system, used an acrophonic notation as early as the second millennium BC. That is, the starting letters of the spoken words became the symbols for the numbers, such as Δ for ten (deka) and H for hundred (hekaton) – much like the Romans would do later. Then, they switched to the use of the actual letters of the alphabet to denote numbers sequentially. The alpha (α) became 1, the beta (β) 2, and so on up to the iota (ι) that became 10. They assigned symbols not only for units but also for tens and higher numbers. The next letter, kapa (κ) became 20, lambda (λ) 30, and so on up to rho (ρ) that stood for 100. They even managed to go much higher with the use of diacritics. An accent under a letter denoted thousands rather than units. The practice of assigning letters to notate numbers continued with the Roman numerals, which became the dominant system in the western world and has had the extraordinary lasting power to still hang around. The Romans assigned symbols for 1 (I), 5 (V), 10 (X), 50 (L), 100 (C), 500 (D), and 1,000 (M). Like most other early systems, compared to how many alternatives could have existed, just a handful of symbols was used to express many numbers. (Expressing numbers higher than 1,000 was quite messy, and fractions were mostly duodecimal, that is, fractions of 12 – but I'll spare you those stories.) Table 1.1 summarizes some notations of numerals. Even though the use of letters further abstracts numbers from their underlying quantities, consider it a step backwards – it caused a lot of damage, as we'll see shortly.

There was still a lot more to come. To be really useful, numbers must be used by people. Besides counting and measuring, this means performing calculations. All early systems managed with record-keeping and other needs but were totally unsuited for calculations by the masses, beyond simple ones that could be performed mentally or by the hand. One big reason for this was that all early systems represented numbers in an additive way. For instance, 203 in Roman is CCIII. It becomes cumbersome to express 1,848 (MDCCCXLVIII), and higher numbers much more so. A simple addition in Roman is daunting. Just try the relatively simple CCLXVIII + LXXXVII, without converting to 268 + 87. Multiplication and division were nightmares. Special techniques and dedicated tools had to be invented. Here we encounter the *table*, the *quipu* (knots) strings of the Incas, the early *rods* and *abacuses* among Greeks and Romans, and the 'modern' abacus in medieval times which became a fixture of life until very recently. Schools were set up to teach the stuff. Instruction lasted for years. A whole industry was developed

**TABLE 1.1**

Symbols used for numbers across different civilizations*

| Modern | 1 | 2 | 3 | 4 | 5 | 6 | 7 | 8 | 9 | 10 | 20 | 100 | 200 | 1000 | 2000 |
|---|---|---|---|---|---|---|---|---|---|---|---|---|---|---|---|
| Ancient Greek | Ι | ΙΙ | ΙΙΙ | ΙΙΙΙ | Γ | ΓΙ | ΓΙΙ | ΓΙΙΙ | ΓΙΙΙΙ | Δ | ΔΔ | Η | ΗΗ | Χ | ΧΧ |
| Later Greek | α | β | γ | δ | ε | ς | ζ | η | θ | ι | κ | ρ | σ | ,α | ,β |
| Roman | Ι | ΙΙ | ΙΙΙ | IV | V | VI | VII | VIII | IX | X | XX | C | CC | M | MM |
| Chinese (rod) vertical | Ι | ΙΙ | ΙΙΙ | ΙΙΙΙ | ΙΙΙΙΙ | Τ | ΤΤ | ΤΤΤ | ΤΤΤΤ | — | = | — | = | — | = |
| Chinese (mandarin) | 一 | 二 | 三 | 四 | 五 | 六 | 七 | 八 | 九 | 十 | 二十 | 一百 | 二百 | 一千 | 两千 |
| Mayan | • | •• | ••• | •••• | ▬ | ▬• | ▬•• | ▬••• | ▬•••• | ═ | ⬮• | ⬮═ | ⬮═ | ⬮⬮═ | ⬮⬮═ |
| Arabic | ١ | ٢ | ٣ | ٤ | ٥ | ٦ | ٧ | ٨ | ٩ | ١٠ | ٢٠ | ١٠٠ | ٢٠٠ | ١٠٠٠ | ٢٠٠٠ |

* The letter stigma (ς) used for 6 in Greek substituted for the earlier digamma (ϝ), but neither exists in the Greek alphabet today. Now, customarily στ is used.

for experts to carry out calculations. Abacists became adored masters. Again, numbers were not for the masses. You needed the equivalent of a PhD to carry out a multiplication.

Even early attempts to name and notate numbers found it practical to use groupings. Unwittingly rather than consciously, these reflected choices for a base and defined the overall structure of a number system. A proper base helps economize on the symbols and the vocabulary needed and, complemented by other necessary conditions later, can facilitate calculations. Any choice of base sooner or later helps discover the power of the *power*, which moves to numbers up the hierarchical scale in a multiplicative manner. In our decimal system, the square of the base moves us from tens to hundreds ($10^2$), the cube of the base to thousands ($10^3$), and so on for higher orders. The equivalent orders in base 5 are 25 ($5^2$) and 125 ($5^3$), while in base 20 they're 400 ($20^2$) and 8,000 ($20^3$). In a complete number system like ours, with base $n$, we really only need $n$ symbols to express all higher numbers without the need to introduce new ones. Using our existing system as a guide, in a system with base 5, 26 would be a three-digit number (101), whereas with base 20 the number 399 would still have only two digits (ΥΥ, if Υ was the symbol for 19). There are advantages and disadvantages to each base. Systems with higher bases create larger numbers easier but the trade-off is many more names and symbols to remember. So, what base to choose and why? This is another problem without a unique solution that reasonable people may disagree. And, did they ever!

All kinds of bases have been tried in history, either in their pure form or as hybrids. Other than notches that used base 1 (unary), base 2 (binary), 5 (quinary), 10 (denary or decimal), 12 (duodecimal), and 20 (vigesimal) were tried. The Sumerians even attempted 60 (sexagesimal), probably through merging the bases of 12 and 5. In principle, such a base would need symbols and names for the first 60 numbers. In practice, to get around such complexity, it was moderated by the intermediation of the 10. Multiple such systems coexisted for centuries. They have left us with many remnants in languages and elsewhere, such as four-twenties for eighty in French, three-twenties-plus-ten for seventy in Danish, the measurement of time (60 min, 60 s) and the circle (360°).

More number systems with different bases were created more recently. The binary system behind digitization and computing only uses 0 and 1. But the low base makes it difficult to express large numbers – a moderate number, such as 76, requires seven digits (1001100). To simplify notation and assist coding, the alphanumeric hexadecimal system is used (base 16), as are systems based on its multiples. The 16 needed digits are 0–9 plus A-F for 10–15. With practice, people can get used to any base and we can move interchangeably from one to another. For instance, denoting the choice of base with a subscript, $76_{10}$ becomes $4C_{16}$ or $1001100_2$. Or $29_{10} = 25_{12} = 1D_{16} = 19_{20}$.

## Towards a Complete Number System

Naming, notation, and the choice of a base in number systems all work in tandem. The lower the base, the fewer the names and symbols needed but the more difficult it is to express larger numbers. Adding symbols to facilitate the representation of higher numbers goes in the wrong direction, as it further complicates calculations. Yet, all cultures did. The Egyptians, the Greeks, the Romans, and others named many high numbers and symbols. The Mayans, who cared about even higher numbers for astronomy, invented more names and elaborate symbols. Such struggles were indicative of something missing, despite many centuries of experience and the reinforcement by parallel advances in math, particularly Euclidean geometry. There was still a long way to go for the completion of our decimal system in a way that we'd recognize it today. Why?

It turns out that the remaining problems could not be resolved satisfactorily until a truly multiplicative and positional number system was developed. A base is a necessary but not a sufficient condition to express all numbers economically. It is through the use of powers of the base that we can move systematically to increasingly higher orders of numbers and fully exploit a multiplicative representation conducive to a place-value system. There, the same symbol represents a different number depending on its pecking order in the hierarchy. For instance, 5 can stand for five, fifty, or five hundred depending on whether it's placed in the units, the tens, or the hundreds. In that context, it also helps if 'meaningless' symbols are used for each number up to the base rather than symbols representing their true underlying quantity. For instance, looking at our two (2) we don't think of two sticks. All these elements would complete the abstraction that is number. Before such a place-value system materialized, notwithstanding some close calls and near-misses, it had to wait for a late arrival. What was missing?

Nothing. *The nothing* to be precise. A big, fat, nothing. The zero. While it had definitely made cameos here and there, it was really out of sight for the longest time, holding everything back. Some number systems flirted with some form of zero but didn't get to know it in its full glory. It appeared in the Babylonians (around 3,000 BC), the Chinese (around year 0), and the Mayans (from the 3rd to 5th centuries AD) but mostly as a placeholder in empty spaces and not as a true number. The Greeks were puzzled by the zero, until Archimedes made a valiant effort to sort it out (and proceeded to use it and estimate the number of grains of sand that would fit into the universe)! Millennia after the invention of number, the Indians[†] nailed it. *Zero* emerged

---

[†]   Thanks to a reviewer for pointing out that 'Indians' really refers to the then inhabitants of the Indian subcontinent.

not only as a placeholder but as the void, the null, to become a real abstract number on its own right and take its rightful place next to the almighty 1, the mystical 3, and the amazing 9.

This is believed to have happened around 500 AD in today's India, and Brahmagupta is credited for using the dot as a symbol after 600 AD. Moreover, it came packaged with a brand new set of nine numerals, which morphed into ours over several centuries. That was the icing on the cake. With a base and the zero, all numbers in a multiplicative and positional system can be expressed by as few symbols as the base-minus-one. In our decimal system the symbols from 0 to 9 suffice. A lot of uncertainty and unnecessary complications were removed from large numbers. Calculations became a breeze. All humans can learn now, and kids will be taught early at school.

The new system started to be used and soon enough it found its way among the Arabs, who had a thriving civilization during a period when the Byzantium was not in its golden age and Europe was in the Dark Ages. Only centuries later it started to move out to Europe, through North Africa and Andalusia. Some people began to be exposed to this new and superior number system but, overall, its welcome was lukewarm at best and its acceptance despairingly slow. Among other things, it upset the establishment. The church had many vested interests at those times, including among powerful abacists, and would treat some *algorists*‡ like heretics and witches under the Inquisition, sending their practice 'underground'.[3]

The Roman numerals and the abacus ruled! Despite its superiority, the wider diffusion of the new system needed additional educational and promotional campaigns that lasted centuries. A big boost was given to it around 1,200 AD by Fibonacci's book but still not nearly enough.[4] The popularization of the new system was helped substantially by the invention of the printing press later in the 15th century. While it wasn't a shoo-in, eventually its popularity matched its superiority and it caught. But even the French revolution had to step in and prop it up, by banning the abacus from schools and government and adopting the metric system.

So, we ended up with the 'Arabic' numerals – which are neither Arabic (as they were invented in the Indian subcontinent) nor are they used by the Arabs to this day! In fact, the Arabs don't want the credit and don't call the numerals Arabic. Our number system enjoys universal acceptance but in many parts of the world it still has a symbiotic relationship with alternative notations. The Chinese gave up their characters in the 1950s but still use indigenous ones, as do Japan, Korea, and much of the Arab world. Also, some other adjustments in the margins remain to be had. We can't quite decide if one thousand should be denoted with a comma (as in North America, 1,000), a dot (as in Europe, 1.000), or adhere to the international standard involving a blank (1 000). And we still have the matter of the billion, which has nine

---

‡ The same word, now *algorithm,* pits sceptics against proponents of *Artificial Intelligence.*

zeros (a thousand millions) but the use of 12 zeros (a million millions) hasn't been quite phased out in some European countries. All in all, compared to all the issues that had to be resolved over the millennia, we can safely call such matters 'small potatoes' today.

To sum up, through a process of tribulation, trial and error, setbacks and successes that lasted tens of thousands of years we ended up with ten symbols, from 0 to 9, that we can use in a decimal, multiplicative, and positional system to express *any* number we can imagine. That's it, that's all.

---

## Is This the End?

We live in an era that even during a walk in the middle of a forest all we need to do is tap on a small screen, see, and talk to any individual on the planet. We can tap on the same screen and see live events happening anywhere at that very moment. We have immediate access to all kinds of information and knowledge. We have gone to the moon, sent probes to the far reaches of the universe, began space travel and aim well beyond. All these are a testament to human ingenuity. The invention of number and the creation of our number system is right up there. With tens of thousands of years in the making, it's an amazing human accomplishment. We are rightly proud of it. Still someone may ask: is this the end? Is the quest over? Is ours the best number system possible?

Now, this is what you may want to call an 'academic argument'. Our number system is what it is. Comparing it with some imaginary ideal is not necessarily a productive thread to follow. Perhaps best to contrast its success with something comparable. The near-universality of our number system among all the billions of humans is much more than we've ever accomplished in language up to now. Despite the disappearance of many thousands of languages and dialects, thousands are still among us. While the Esperanto idea may have been 'logical' and appealing, it hasn't worked. This makes things interesting, and we value the diversity. But can anyone today fathom the same diversity in number systems, when the mere coexistence of metric and imperial units poses challenges?

Numbers are our brainchild, not the products of our senses. They represent an amazing invention of our minds, not a discovery of something that was there if only we searched hard enough. In his monumental work, *The Universal History of Numbers*, Georges Ifrah says that *"The invention of numbers was not a smooth, logical sequence"* and that *"Logic was not the guiding light of the history of number-systems"*. As to whether or not our system is the best possible, I'll share his answer in a moment that captures both his enthusiasm and excitement: *"This really is the end. Our positional number-system is perfect and complete, because it is as economical in symbols as can be and can represent any number, however large. Also, as we have seen, it is the most efficacious in that it allows everyone to do arithmetic."*

Having said that, in the spirit of never-say-never, human curiosity and ingenuity know no end. Betting against any further breakthroughs is never safe. If asked to devise a superior number system, brilliant people in our midst can surely give it a shot. They may recommend, say, a system with base 12, which has more divisors than base 10 (2, 3, 4, and 6 rather than only 2 and 5) and is friendlier to fractions. Or they may recommend a prime number base, say 11, something that would eliminate duplication in expressing fractions and take care of some other annoyances. As we've already seen, new number systems have already been devised for computing. Such developments will undoubtedly continue. In parallel, in a more abstract realm, mathematicians have long gone beyond the early number. Inventions include *negative* numbers, *perfect* and *weird* numbers, *complex* and *imaginary* numbers for two dimensions, and their extension to *quaternions and octonions* for four and eight dimensions. There's always been a fuss over irrational numbers, which can be algebraic (e.g., solutions of some polynomial equation) or transcendental. Mystical properties have been assigned to prime numbers, the 'divine proportion' of the *golden ratio* ($\varphi = 1.618...$) and, of course, the ever-elusive and mysterious transcendentals e ($2.718...$) and $\pi$ ($3.141...$) – the latter making it impossible to square the circle.[5]

The origins of numbers can be traced directly to today's numerical data. In that sense, we owe a debt of gratitude to all those who toiled over the millennia and contributed their proverbial stone to this magnificent edifice. From our early ancestor, who carved notches on a baboon bone, to the scribe in ancient Egypt who recorded the annual crop, to the Mayan who measured the year with such precision, and the Indian who invented the zero before Florence Nightingale recorded and visually presented statistics for the cause of death of British soldiers in Crimea. The chain continues today with the statistician who strives to provide data in support of the UN's *sustainable development goals* and the 'big data' researcher. Increasingly, such work keeps many meaningfully occupied and continues to feed our data, which have been called the new fuel. Number is to have and to hold.

## Take-Aways

Other than a nice story what take-aways are to be had from this detour to the birth place of number and the development of our number system? What useful insights can we carry along as we go through this book? I'll touch on a few below, and I hope you can think of more as you mull things over.

*Numbers and communication:* A good invention amplifies our natural abilities. A lever helps lift more weight than our own strength, binoculars allow us to see farther than our eyes can, a computer can perform calculations much faster than we can. Numbers too represent an invention that extends our powers to quantify beyond vagueness, something particularly evident in counting and measuring. We look at a 'small' distance but don't know it's 2.3 m, an 'average' room but we can't tell it's 15.6 m$^2$, a 'tall and big' individual but don't know he's 1.90 m and 96 kg, two 'far away' stars but can't tell the difference between 17 and 23 light years. Old-style directions like *"go straight for about 300 metres, turn left at the gas station, drive two blocks before you make a right (you'll see a corner store), keep going until a tall cedar fence that you can't miss, then one more right and you'll be here"* can be stressful – and have led to mishaps. Contrast this to *"25 Someplace Road"* and a map. In communications, numbers accomplish the equivalent of this, providing both the address and the map. They convey information more succinctly, more accurately, and more efficiently. It's important to comprehend numbers at a deeper level, as quantities that exist beyond conventional numerals or measurement units that pop in our minds due to our years of conditioning. Numbers are a true universal language – and a very good one at that. Yet, like all languages, it can be used with different degrees of sophistication. Slang and poetry can live side by side.

*Numbers and human development:* While an extraordinary accomplishment of the human mind, numbers historically advanced through the gathering and use of quantitative data or *statistics*. The word originally derives from *state* and the underlying measured activities are characteristics of organized societies. There's nothing accidental in that all progress is associated with 'civilizations'. Our digital era, the product of an unprecedented technological explosion, is awash with data. The more diversified and complex the affairs of a collectivity become, the bigger the need for numbers. The 'one-two-many' cultures corroborate this too, if more ammunition was needed. In the jungle, as we saw earlier, our response to the threat posed by a pack of wolves won't be determined by their exact number. In well-functioning societies, with thriving economies, data are not merely products of curiosity. They're indispensable to quantify transactions, record various happenings, reveal (in)convenient truths, monitor accountability, and so much more. To this day, statistics and the capacity to produce them are easier to spot in countries with more developed and open economies. Most international studies, in any field imaginable, stumble on to the dearth of data in less advanced countries. There is also a negative correlation between statistics and the degree of authoritarianism in a society. Look for data to be among the first casualties in regimes antithetical to transparency.

*Addressing innumeracy:* Despite the millennia, evolution has still not given us yet an organ for numbers. (Noam Chomsky notes that a mutation at some point did give us some genes for language!) I don't know if it ever will, but as a prime candidate I don't feel anything coming. So, remember our innate inability to discern quantities beyond four and that numbers don't come naturally to people. We have to patiently recreate in every new human the eureka moment of our distant past. Our ancestor's campaign for numeracy continues to this day, and this book is proof. Moreover, the availability of data and our understanding of them must be matched by our ability to manipulate them. Think of that next time you're facing a spreadsheet or some Python code.

*For better or for worse*: Most of my professional life I experienced firsthand the reactions of people who needed some data for something they cared, here and now. When such data didn't exist, weren't nearly enough or exactly what they thought they should be, the element of surprise was frequently over the top, occasionally amusing, and always genuinely painful. You could tell that people were under the spell of a sincere impression that someone, somewhere, somehow measures *anything and everything* they may wake up desiring. More recently, though, I found myself surprised with experiences of the opposite. Often times I wondered how come some data exist. One such example was when I found out from Google the number of places I had visited last July, amidst many more details. Good that someone knows – in case I have to answer a survey on my whereabouts. We'll discuss all that and much more in the chapters ahead.

## For the Love of Numbers

Numbers are wonderful. You can take some of them or all of them, toss them in the air, and guarantee that a number will be formed any which way they land, every time. If you attempt this with letters, you'll get gibberish more often than not. Languages evolve continuously and every year there are new words – but no new numbers. And this is not because there are 26 letters (in the Latin alphabet) but only ten numerals. Numbers have no end, yet they're orderly. Their structure is vast, yet not messy but elegant. This stuff is awesome! I can see why the order within their endlessness, their symmetry, and self-contained 'completeness' can fascinate. They can even provide an escapism from the imperfections of the real world. Indeed, some became outright obsessed with the beauty of numbers, mystified by the 'truth' and the promise they embody – to the point of torture. Early on, Pythagoras and his disciples created cult-like schools for the study of numbers, among many other things to be fair. 'Everything is number' or can be expressed in numbers, from the sounds of music to the existence of the earth and the movement of the universe. Magical properties were assigned to numbers. Such believers became so wound up in their own world that some amongst them would never accept that their beautiful numbers could possibly be subject to any 'flaw'. Going beyond their adored integers and fractions, and faced with inescapable proof of the existence of irrational numbers (such as $\sqrt{2}$ or $\pi$), in modern lingo they flipped, they lost it, went berserk. They just couldn't swallow such a blow. This 'error' on the part of the Supreme Architect spoils the harmony of the world, it cannot be spoken. It had to be kept a secret, an *unmentionable* anathema. Accounts have it that someone turned suicidal.

Numbers are wonderful. Yes, they command respect. But we cannot allow our respect for numbers to blind us. They're our creation, not our God. In the form of numerical data, they're frequently the harder parts of our evidence. Despite this, or especially because of this, questioning numbers is to be expected. Believe but verify. Scrutinize any number you see. Nowadays, with more and faster information flows than ever, we have to put up with misinformation, disinformation, and lies. What we're experiencing is qualitatively different from what was implied by the much-quoted *"lies, damn lies, and statistics"*, credited to Disraeli. Our response to the old and the new threat, though, has common elements. Good numbers help weed out bad ones. Critical thinking and questioning based on know-how are a must for fact-checking. So, as part of our ongoing journey, every chapter will end with some thoughts and tips on what to look for or how to go about it. The intent is to deliver them in an opportunistic and light-hearted manner rather than the stern way of an exhaustive checklist or a manual.

## Fact-Checking Tips

*Homogeneity, counting, and interpretation of statistics:* An early step in the invention of number was the realization of likeness among units, animate or inanimate. Our ancestor carved notches on a bone by abstracting the differences among units. This carries through to our times when we affix a count to taxonomical groupings that represent aggregations of units subject to considerable diversity, something exemplified in the widespread use of classifications. Remember this next time you see statistics for households, businesses, or houses. A headline that *"Total business revenue increased by 10% last year"* may well be accurate, yet only part of the story. *Businesses* include multinational behemoths, small and medium enterprises (SMEs), and micro-enterprises. Very large businesses typically account for a tiny fraction of all units but for the bulk of revenues. Units in *restaurants* are more homogeneous than those in *businesses* and less homogeneous than those in *pizzerias*. Depending on the actual statistic reported, unit size, sector, geography, or some other variable may matter in different ways. Decomposing by size, for instance, could reveal findings that would modify the headline to *"Total business revenue increased by 10% last year despite that revenues declined in 70% of businesses"*. A more flamboyant writer might even express it without numbers as *"Big business thrives, everyone else suffers"*. The story may still be different for businesses with less than ten employees. Even when true, top-level messages may be much more nuanced when checked against granular data. In the very least, this kind of *robustness* of the reported data affects our understanding of what's happening and is instructive of how to internalize the message regardless of how the source presented it.

*Large numbers and vagueness:* Large numbers made the ancients stumble, on multiple fronts. Naming, writing, and calculating were all challenging. The Greeks used *myriad* for 10,000, and the Romans would say *ten-hundred-thousands* for a million. The Indians gave names to an astonishing number of very large numbers, up to $10^{421}$, enough to fill tens of pages. Today, we're more at ease with some large numbers but our relationship remains challenging.

Even highly educated people trip, and feel compelled to spell *billion* with a *b* – and this is not a large number today. Efforts to gain familiarity through analogies based on orders of magnitude help, up to a point. Explaining the relative differences of the extra three zeros that separate a thousand from a million, then a billion, and a trillion as equivalent to moving from 17 min to 11½ days to 32 years and to 32,000 years probably helps retention! A ball of a certain size held at a certain distance from an orange may well help capture the proportional distance between the earth and the moon. Perhaps the volume of information contained in a drawer rather than a thousand libraries helps wrap our heads around the difference between gigabytes ($1,000^3$ bytes) and zettabytes ($1,000^7$ bytes). We may or may not know that Google derives from googol, 1 followed by 100 zeros ($10^{100}$), but certainly few comprehend the meaning of the gap between an octillion ($10^{27}$) and an octodecillion ($10^{57}$), and no one can claim a real appreciation of the difference in distance between the Andromeda Galaxy (about 2.5 million light years away) and the Great Attractor (somewhere between 150 and 250 million light years away).[6] At those scales, comprehensible orders of magnitude collapse and all-too-often we resort to *astronomical (sic)* differences. We have even created words like *gazillion,* precisely to bypass this inability. Very large numbers represent a classic case in which the audience has a hard time spotting wrong in the event we're taken for a ride. Don't hesitate to use any means at your disposal, other than numbers. When exact counts are not crucial and the objective is conducive to orders of magnitude, a picture nowadays also extends the natural ability of our eye and will settle issues of comparative crowd size in a given locale at two different times. There's no excuse to confuse a million, a billion, or a trillion today but when you feel that numbers beyond our comprehension need checking, rather than attempting to become an expert, remember that we're only a click or two away from finding out what we must. Accessing expert resources when needed is often a prudent course of action.

*Absence of numbers:* You may encounter cases in which the absence of numbers is as suspect as the use of the wrong numbers. Substituting vagueness for numbers can come from any source and can be accidental or intentional, the result of research laziness or malice. Always beware of stuff such as "*Some say that most people prefer apples as a fruit*". Words like *some, many, most* are frequently used when the number is not known or is too weak to support the argument – which is made anyways. Particularly when you hear *some say* ask yourself "*who's that guy*", as the chances that it was one person are high. You'll frequently come across unnecessary confusion between *majority* (50% + 1) and *plurality* when the set of choices extends beyond two. Was the apple chosen against an orange or among an orange, a banana, a kiwi, a peach, and a bunch of grapes and strawberries? Could it be that *most* was 16%, with three other choices at 15%? Some preference! Substituting vagueness for numbers is okay with friends over coffee but not in scientific results, policy work, or business decisions.

*Spot a charlatan:* All kinds of ridiculous things have occurred in the name of numbers too, including assigning to them mystical properties – even gender. Odd numbers were male, even numbers female. *One* was God, *seven* was lucky, *ten* was perfection. Thankfully, most such things have been phased out of existence. Sadly, something still lingers around causing lasting damage.

The arrival of the Phoenician alphabet and its different adaptations led to the proliferation of written languages. While this was another magnificent human accomplishment, it cast a dark shadow on numbers. Our number system was not sufficiently advanced yet, and letters were used as part of the efforts to sort out notation. I told you earlier that this was a step back-wards in the evolution of numbers, and a terrible idea. Not only it impeded and delayed the development of the number system but eventually it inflicted worse harm. It gave rise to numerology, which has as much to do with numbers as astrology has to do with astronomy. Rooted in the ancient Chaldeans in Babylon, this practice was taken up by the Hebrews, the Greeks, the Arabs, and Europeans with Latin alphabets. No one escaped. When letters denote numbers, words inevitably sum up to a numerical value. But any cor-respondence between numbers and letters is nonsensical. The Latin alphabet has 26 letters, with z being the last. The Greek alphabet has 24 letters, with z being the sixth, it's using $\mu\pi$ in lieu of $b$, and contains the *omega*. Even if there was an identical word, when written in the two alphabets what number should one expect? This reality check alone should have been plenty to stop the nonsense in its tracks before many more alphabets got involved. It's par-ticularly rich to see that other culprits were languages that only sparingly use vowels in writing, like Hebrew and Arabic with 22 and 28 letters, respect-ively. Yet, things got seriously out of control leading to a dog's breakfast. Except for some relatively innocuous *anagrams, chronograms,* or *palindromes,* it really led people astray to utter nonsense – even the borders of sorcery. Declarations of sacred and diabolic words emerged. Numerology created strands like *gematria* (Jewish alphanumeric codes), *isopsephy, arithmancy,* even *abracadabra* whose 'mysticism' finds the 'key number of a first name' through a series of triangles containing letters that add up to 365. With much fiddling, the protestant Michael Stifel made sure that Pope Leo X's number was 666, making him the anti-Christ. Not taking it standing idle, and also with a lot of acrobatics, the catholic theologian Peter Bungus made sure that 666 really belonged to Martin Luther. All kinds of such nonsense proliferated, wreaking havoc along the way.

Numerology became extremely widespread. Countless books have been written on the subject and the amounts of energy wasted are truly regrettable. You can think of all that as the first major case of disinformation, over-the-top and in-your-face gaslighting, and conspiracy theories gone wild! Alas, it's still absorbing valuable oxygen. This is from a recent visit to a numerology website (I only removed a commercial name, not to identify it): *"These numbers are used to offer insight to personality, future events, and even life's greater purpose"* immediately followed with *"Check out this month's [name] Beauty Box, which is packed with products hand-picked by our editors – all for only $15"*.

*Know a game when you see it:* We use numbers to make sense of our economies, organize our societies, do science, go to the moon, and much more. While there is no shortage of very serious uses, I never said we can't use numbers to play and have fun too! In fact, numbers are a lot of fun in games. Some are entirely based on them (Sudoku), others can be derived on the fly because of properties embedded in our particular decimal system. All kinds of neat relationships among numbers have been discovered. From shortcuts to multiplication (how to square numbers ending in 5) to finding the day behind any date ever and a whole array of cute mental tricks. A particular one is that if you keep adding up the digits of any product of 9 you always end up with 9. This added stock to the mystical properties of 9 but is, of course, strictly a product of our particular number system. All such stuff has no independent existence and would vanish in thin air under a number system with an alternative base. Naturally, new ones would then emerge – but that's exactly the point. Similar to how we distinguish cardinal numbers that capture quantities from ordinal numbers that indicate rank in sequences, dissociating the essence of the underlying quantification from peculiarities in our familiar number system is important to avoid game-like pitfalls in our fact-checking activities.

---

## Notes

1 The material included images, music, and greetings in many languages. Mathematical images sent with Voyageur 1 and 2 can be seen at: NASA, Jet Propulsion Laboratory, California Institute of Technology, "What are the contents of the Golden Record?" Image is called Mathematical Definitions, by ©Frank Drake https://voyager.jpl.nasa.gov/golden-record/whats-on-the-record/

2  Georges Ifrah. *The Universal History of Numbers: From Prehistory to the Invention of the Computer.* Wiley, 2000.
3  Against the prevailing winds, Gerbert of Aurillac (a monastery in France), who in 999 became Pope Sylvester II, was an early supporter of the new system. It was called *algorism* from a bad Latinization of the name of al-Khwarizmi, a well-known Persian scholar from Khwarazm (modern day Uzbekistan), whose 9th century book *On the Calculation with Hindu Numerals* was translated in Europe in the 10th century.
4  Leonardo Fibonacci had lived in North Africa before he returned to Italy. His 1202 book *Liber Abaci* (*The Book of Calculations*) exposed the new number system to many in Europe but also presented the *Fibonacci sequence*, for which he became famous. The sequence is connected to the golden ratio φ.
5  Transcendental numbers are not only irrational but also they don't satisfy any algebraic equation. More than 2,500 years after the Pythagoreans, the mathematician Leopold Kronecker declared that π doesn't really exist because we don't know all its terms! Tony Crilly, in his book *The Big Questions: Mathematics* (Quercus Publishing), 2011, quoted Kronecker's famous words: *"The natural numbers come from God, everything else is the work of man"*. Yet, in 2014, Daniel Tammet, who's diagnosed with Asperger and savant syndromes, recited live 22,514 digits of π! See also his book: Daniel, Tammet. *Thinking in Numbers: On Life, Love, Meaning and Math.* Back Bay Books, Little, Brown and Company, 2012. The mathematician Mario Livio devoted an entire book to the golden ratio, and exalted π. (Mario Livio. *The Golden Ratio: The Story of Phi, The World's Most Astonishing Number.* Broadway Books, 2002.)
6  Wikipedia. "Light-year." Updated April 30, 2022. https://en.wikipedia.org/wiki/Light-year

# 2

## Measuring, With Instruments

With numbers at hand, we can go beyond the relative orders of magnitude that our natural abilities register. Not only can we count, we can measure too! In fact, quantification of our world becomes more of a mechanical and operational exercise. All we need is to make up some units and we're on our way. Or is there more to it?

### Units and Instruments

Our physical world is three-dimensional. Luckily, length, width, and height (or depth) are all distances. The same unit that can measure the proverbial distance from A to B can measure the height of mountains. Even better, now that we *really* understand number, any unit will do. Surely, a shorter unit will produce a larger number than a smaller one. But whatever nominal number we end up with is not important. What matters is the value of any measurement against the unit. The same is true for weight, volume, and everything else.

Sure enough, from early on different people came up with a variety of units. The ancient *cubit*, roughly an arm's length, helped the construction of the pyramids. Relying again on human physiology, the *foot*, the *pace*, and the *hand* were among those used for distance; the *mina*, the *shekel*, and the *libra (pound)* were used for weight. Units for area, volume, and more were added to the mix. People grew confident. Why stop here? The curious child takes over. Why can't we venture beyond the visible, to 'things' we experience, things we know *are* there. Wouldn't it be nice to measure temperature and humidity? What about the winds? And we absolutely feel compelled to measure time. Out come the *solar clock*, the *clepsydra*, and the *hourglass*. Before we know it, the stars are in play. What's happening out there? But there was still time before we discovered so much more to measure, like gravity, electromagnetic fields, black holes, or happiness. For now, let's look at the near neighbourhood.

Early units of measurement match immediate needs and, with enough use, they evolve. Take distance as an example. Initially, the choice of unit will be driven by the relative length we try to measure. A foot may be a good choice

DOI: 10.1201/9781003330806-2

for a bedroom but the pace is more convenient for a plot of land. A metre doesn't take us far on a stretch of highway, a kilometre works better. Some multiple of the unit is more suitable when geodesic distances are concerned, such as the circumference of the earth. And when our eyes are set on the stars, thousands of kilometres ($10m^4$) don't cut it. Then, a metre is too long for the length of a pebble. Rather than sweating the complications from many different units, just scale one – up and down. So, multiples and fractions of units were devised, both for convenience and precision.[1]

Yet, there's a difference between counting and measuring in as much as the latter requires an implement to represent the unit. But unlike the reliance on a fixed-length stick or a simple ruler for distance, measuring weight requires a scale – a more elaborate construct. Other measurements need even more specialized instruments. The realization that the precision of the measurement depends on the quality of the instrument had practical implications. Instrumentation became intertwined with units, and figures prominently in the evolution of measurements. The history of making timepieces alone can captivate and keep someone busy for a long time. Improvements in instrumentation rose to another level with advances in science and engineering, and continue to this day with levels of precision at levels incomprehensible to most of us. (Do check out the latest atomic clock).[2] For millennia, however, the world had to make do with what it had. You could understand if in the old days someone asked *"where did you get your cubit"* or *"my scale or yours?"* This raised the need for standards.

### Standards

As the need for more measurements emerged to serve agriculture, land property rights or construction local solutions proliferated and there were many variations of units everywhere. With expanding commerce and cross-cultural trade, it became increasingly evident that common units would be a good idea. Conversions between units weren't fun, particularly with the limited capacity for calculations at that time. Errors would be avoided, cheating and disputes would be managed better, communication would be easier. This is a practical matter after all, not one for egos. Nonetheless, unlike numbers, we're still trying on this one. One more example of people making things messy before we decide to get our act together and clean up.

The process started early. An example was a metal rod equal to a cubit, modelled after the arm of the Pharaoh. It was kept tight and safeguarded as the standard but didn't quite succeed. History is replete with non-compliance, and different cubits. Serious attempts at standardization had to wait for a very long time, until the advancement of science and industry. Calls for a 'stable and coherent system' with precisely defined 'permanent and unmodified standards' picked up momentum only in the 17th century AD. The best collective effort led to the development of a metric system but its (partial) adoption had to wait until the end of the 18th century, when the French revolution put its muscle behind it.

Although not all proposed standards were adopted (i.e., metric time was not), and not all countries signed off, progress was definitely made. Very slowly, the metric system has taken hold in most of the world. Unfortunately, the British opted out – as did the United States (US) with its heavy influence on science and world affairs. (In 1790, Thomas Jefferson had proposed to Congress that the US adopt the metric system.) Rather than metres, kilograms, and litres the imperial system relies on feet, pounds, and gallons. As well, in the US the Fahrenheit scale is used for temperature rather than the Celsius scale, which is used in most other places – and neither of them is the international standard! In Canada, even decades after the adoption of the metric system, the use of feet and pounds is still prevalent in everyday life. Today, there are seven *fundamental physical quantities* in the 'international' metric system: metre (distance), kilogram (mass), second (time), kelvin (temperature), ampere (electric current), mole (amount of substance), and candela (luminous intensity).

The early definitions of standards were complex enough for then, but they are totally unfit for mere mortals today.[3] Then, again, this won't affect the metre tape or the digital scale we'll buy at the hardware store. The bottom line is that, unlike numbers, we have yet to arrive at a global system. The more ingrained units become in the functioning of advanced societies, and the more embedded in the archives of accumulated knowledge, the more difficult it becomes to part company. But the story of *standards* in a more general sense is a much larger discussion, which doesn't stop at measurements. Just think of trains that can't cross borders because of different track widths or the number of power converters in your carry-on bag when you travel. Thankfully, the internet dealt with this matter upfront and was built on a common *protocol* – imagine otherwise!

## Interrelationships Among Units

Standard units or not, yours or mine, how do we actually measure? How do we go about negotiating a kilometre? While we can still buy a 25-m tape, we can't buy a kilometre-long tape. And what if we could? How do we cope with a tonne? Things like that presented challenges over the millennia. But no longer. Today, we don't need any of that. Sometimes even the *idea* of a unit is enough. Science and technology to the rescue.

From early on, measurements were given a boost by math. This was particularly true for distances, areas, and volumes thanks in large part to Euclidian geometry and subsequent advances in trigonometry and stereometry. Those were the days of a very fruitful partnership between math and numbers, when one helped the other. (That union has gone sour long since.) Whether squares or triangles, circles or cylinders, pyramids or octahedra, math greatly facilitated their measurement. We only needed to measure one side of a square, two sides of a right-angled triangle, or two sides and an angle of *any* triangle to know everything else. We can calculate their perimeters and their areas without actually *measuring* them directly. The volume of liquid in an octahedron container can be computed in a similar fashion. The height of a tree can be measured from far away.

Later, science added major breakthroughs by unveiling interrelationships among units, such as $d = t \times s$ (distance = time × speed). With the speed constant, time will exactly measure the distance – with a clock rather than a metre. The odometer was not far behind. We now have another abstraction, moving from the standard unit of a measuring instrument to the unit as a mere reference amount. With parallel advances in navigation and mapping we could estimate large distances better than ever, including the circumference of the earth and the distance to the moon. And we kept improving. We measured the speed of light and decided that a *light year* suits better as a unit of distance. It sure sounds a lot better than 9.46 trillion km ($9.46 \times 10^{12}$) or 5.88 trillion miles ($5.88 \times 10^{12}$), doesn't it? It wasn't long before we used *parsecs* (3.26 light years) for even more vast distances in deep space. Improvements in measurement were brought about by better instrumentation as much as by conceptual advances. Application of similar principles has brought to our local hardware store inexpensive electronic devices to measure easily and accurately all kinds of things in our daily lives. They can even be downloaded on our smartphones.

Using proxy units as convenient shortcuts can confer advantages but the context always matters. Saying that *"I live half an hour outside of Ottawa"* is more common than giving a kilometric distance from the city's centre. When it comes to travel, a lot of unnecessary confusion exists that leads to waste of time. For instance, depending on the exact origin and destination, door-to-door travel between two adjacent cities can differ by tens of kilometres. Frequently, distances are measured between downtown cores or based on fixed points, like train stations. Such is the case for the 259 km between gare du midi in Paris and gare du Nord in Brussels, which takes an average of 1 h 30 min by train. Exact longitude and latitude coordinates are used for 'as the crow flies' estimates. Such estimates can be fairly close to, or very far apart from, 'real' distances. How winding the road network is has a lot to do with the actual driving distance. When time and cost are at issue, speed limits, tolls, hairpin turns in mountainous terrains, and possible ferry crossings matter quite a bit. These days, a GPS-based online search will indicate not only the shortest but also the fastest route from your house to a place of interest. Moreover, it'll indicate the time driving, biking, or walking.

## Expansion of Measurements and Instruments

From the physical that we see to the intangible that we feel, from what we didn't know existed to what we suspected might be there, from the needed to the nice-to-have, we've come a long way. We attach numbers to temperature, humidity, winds, their combinations (humidex and windchill), barometric pressure, electromagnetic fields, and data flows. We make quantities out of all sorts of occurrences or phenomena. We devised numerical scales to quantify earthquakes (Richter and Mercalli), wind forces (Beaufort), and more. Sonars use acoustic waves to measure the depth of the oceans, lidar uses remote sensing to measure the shape of the earth in 3D. Among the speed of sound and of light, gravity and the Planck constant, we quantified fundamental physical phenomena that encapsulate our understanding of our world and its place in the universe. We got inside our bodies. We measure our pulse, heartbeat, systolic and diastolic blood pressure, sugar levels, good and bad cholesterol. Are we there yet?

Not only are we now measuring a dizzying number of things but we have devised a vast number of instruments. (Notice the use of *dizzying* and *vast* before *number*? Translation: I don't know and am not inclined to ~~measure~~ – oops, count.) A simple search will put your head into a frantic spin. Here are a few you can likely make out – and only among those ending in meter: altimeter, anemometer, barometer, electrometer, fathometer, manometer, odometer, rodometer, trumeter, seismometer, spectrometer, tachometer, voltmeter. Particularly in new developments, applied science with its instrumentation has *de facto* rendered the issue of standards moot.

---

## Superior Data?

Except for the intrinsic significance of measurements in the world of numbers, there's an additional reason for the discussion above. Much has been made of the distinction between the investigation of the physical world through science and the study of human affairs through 'softer' sciences, fields, or disciplines. Surely, well-known differences exist, not the least of which is the issue of the reproducibility of results under controlled, lab-type experimentation. Part and parcel of that is the heavy reliance of 'hard science' on instrumentation for better data, from the subatomic level to the cosmic. Data needed for proof of the hypothesized existence of *black holes* or *strings* require even colossal structures like gigantic telescopes, spacecraft, and particle colliders. The Large Hadron Collider at CERN, where the *God particle* (Higgs boson) was discovered, lies in a tunnel as deep as 175 m and has a circumference of 27 km, spanning France and Switzerland. In that context, terminology that refers to surveys as *measurement instruments* is purely metaphorical but language has a way to borrow opportunistically (learning through osmosis). While looking up to the *big brother* is no shame, perfecting instruments through advanced manufacturing or new materials is not in the cards for improved measurements in social sciences. All the computers in the world, while greatly help capture, process, and record statistical data, cannot become measuring instruments themselves. They will not improve how we 'measure' the notion of output (GDP), the price level (CPI), or the rate of unemployment. New and improved data sources, as well as innovative methods, will. The point is that researchers in the socio-economic realm can also postulate constructs in their minds freely, unencumbered from the impossibility of having some *inflationometer*. They will try to quantify them with any means at their disposal. A more apt term for such quantification would be *estimation*. In the final analysis, data are data regardless of their source. So long as they're present, statistical and analytical processes will also be present in earnest.

Accuracy, whose attributes include precision, is a precious and sought-after quality in measurements. Nothing worse than no data but bad data. Echoing the previous discussion, oftentimes you'll encounter the perception that data derived from physical measures are accurate, stable, and free from baggage like margins of error or confidence intervals. They're therefore far more reliable than socio-economic data. You'll also come across that the choice of unit affects the accuracy of measurements. More specifically, the smaller the unit, the more precise the measurement. By stepping on such assertions, it's not too early to dig a little deeper and start examining some crucial aspects of statistical work – which we'll meet again later. We'll approach such matters both theoretically and empirically.

## Games of the Mind

Clever arguments can cherry-pick cases in which even simple physical measures cannot be perfectly exact, even with the best of instruments. First, remember the Pythagoreans? The existence of irrational numbers drove them to insanity. The Pythagorean theorem itself proved conclusively that the length of the hypotenuse of a right-angled triangle whose other two sides have a length of 1 can never be exact ($\sqrt{2}$). Similar is the case involving the circumference of a circle. No matter how carefully we measure all around, with the tiniest linear unit imaginable, we can't get as good a measure as $2\pi r$. Even the most accurate measurement of r (the radius) won't help given the presence of $\pi$. Practically, should the hypotenuse develop an inferiority complex about its 'normality' vis-à-vis the other two sides? Will any of that put a dent on your design plans involving a circle?

Second, no measurement's expression is constrained to multiples of full units. Choosing a smaller unit to improve the precision of a measurement is one and the same as choosing a smaller subdivision of any unit. In that trivial sense, a smaller unit cannot be part of any answer to the issue of improved precision – which, I suspect, relates to the measurement of the contour of irregular surfaces with the classic example being the length of a coastline. In the case of distance, pushing the issue of a smaller unit to the limit soon bumps up against the inability to define the smallest conceivable distance between points A and B, as we can't define the minimum possible 'smallness' of the abstract and quantity-free Euclidean *point*. Such inability, however, hasn't been a weakness in any practical applications and uses for millennia. So, if not the unit, what then affects the accuracy of the measurement of a coastline?

Ignoring the obvious shapeshifting from the back-and-forth of the waves and the movement of the sand, even in static form a coastline displays fractal properties from chaos theory.* Truth is that under fractal conditions, the more closely you look the more details you see. It's not just a matter of measuring all around a small protruding pebble here and there but that, under a microscope, what seemed to be a smooth pebble is not – it's quite rugged with its own protrusions, each of which has even more of its own. We'd need to measure all around them too, in a process with no end. So, as the scale on which the measurement applies becomes more granular, tinier, and approaches zero, the length increases – with infinity as its limit. Rather than the coastline of Africa, all that can happen within a small cove only. Or on a head of broccoli. But it's the *scale* that causes that, not the *unit* used. In other words, any measurement obtained is a function of the specific resolution chosen. Never mind how infinitesimal the unit, so long as a higher resolution reveals more coastline, we undercount. However, such thinking is equivalent to allowing the discovery that most of the space in an atom is empty to shake your faith in the existence of solids – until you hit your toe somewhere.

Interesting for the mind as such things may be, we need to keep moving. Other than a migraine, the useful takeaway is some lingering and useful doubt on the absolute precision of any measurement – lest we become overly enamoured and cocky over some particularly good numbers. In real life we'll get a fairly accurate measure even with a humble metre. The practical extent of our problem may be choosing between a measure of two millimetres or two-sixteenths of an inch. A bigger lesson, though, is to start thinking seriously about the relationship between the needed accuracy of our data and

---

* In fractals, a shape retains the same pattern of irregularity at different scales with no end in sight. The magnified image of the structure of an atom, with the electrons hovering around a nucleus, is indistinguishable from the heavily shrank image of some exoplanets and their moons revolving around a sun. Fractals can be produced by something as deceptively simple as $x_{t+1} = kx_t(1 - x_t)$, depending on the values used.

their intended use. This is frequently misunderstood and causes a fair amount of friction. When data accuracy is at play, we must always have the 'so what' question in mind. Granted, it's overused by middle management but it still helps ground us to the matters that our data aim to illuminate.

## Empirical Matters

Strictly speaking, it's not true that physical measurements are not subject to a margin of error. Such assertions probably stem from equating *error* with the *margin of error* reported in opinion polls or other surveys. But there's an easy way out of this. What is commonly reported represents the sampling error of such exercises and is really the *margin of the known error* (*knowable* would be even better, if I were picky). Reported results from good surveys should include metadata that make explicit reference to both sampling and non-sampling errors. The margin of error should be the sum of the two. Typically, however, non-sampling errors are not quantified even though they can conceivably exceed the sampling errors. While physical measurements are not subject to sampling errors, they are subject to all other. There's always something, from imperfections in the measuring instrument to human error. In the same vein, reproducing identical results across repeat experiments is rarely the case even under controlled lab conditions. A perfectly 'fair' die is an imaginary concept, much like a point in Euclidean geometry. Practically, there's always something to affect the accuracy of our numbers. Again, what matters is whether they're fit for their intended use.

At times the allure of physical measurements stems from the fact that they may be fixed – not exactly a nuanced distinction from being accurate. Surely, we can get accurate measurements of a fixed distance or weight, such as the distance between two coordinates as the crow flies, the circumference of the earth, or the weight of a boulder. But even better, such measures have the additional advantage that we can park them. Being time-invariant, there's no frequency to contend with. We measure once, record, and save forever. No matter how daunting the task of an accurate measurement may be, once accomplished we move on. Fixed measures are also useful to anchor or facilitate subsequent measurements. The cargo weight on a truck can be found by weighing the loaded truck and subtracting the truck's known weight.[†] Similarly for a pair of boots or a piece of clothing vis-à-vis our weight on a scale. We can get an accurate measure without being shoeless or naked. As human affairs are not governed by physical laws, socio-economic data are mainly estimates and not privileged to such fixities. Whenever such a chance appears, though, you grab it, and you call it a good day at the office. Shelf

---

[†] Even then, obsession with accuracy could open the door to how inflated the tires are or if a cell phone was left inside.

prices and interest rates observed at a point in time, or exchange rates at the close of the trading day, are simply recorded. No margin of error, no need to revisit – or revise.

## Shifting Interests

In reality, you'll encounter inconsistent numbers even for what are meant to be unique physical measures. Such is the world of numbers! You'll have to put up with multiple sources, slightly different approaches, and other 'details' and twists. To see also how physical measurements can quickly get mixed with socio-economic data and how units of interest morph to match different needs, think of highway networks that constitute a critical path of a country's physical infrastructure. It's useful to know that the Trans-Canada Highway runs for 7,821 km from St. John's in Newfoundland to Victoria in British Columbia. While you have faith that there's a road available all the way from the Atlantic to the Pacific, the exact linear distance may not be of much interest – and you'll come across many different ones when you really need a number![4] But you'll care a lot more about an upcoming trip from Montreal to Toronto. The distance of 542 km (again one of several) will help you plan the trip, in terms of gas cost and time. As you search for the distance you may come across something like the existence of 2,000 km of two-lane equivalent highway between the two cities. What are you supposed to make of that length?

The linear distance to get there is of interest to you. But then, a highway runs in two directions. The linear distance doubles, there are 542 km both ways. Then, typically, highways have two lanes in each direction. But until you're out of the island of Montreal the highway has three or four lanes, as it does 'an hour or so' outside of Toronto. For a long stretch before downtown, there may be more lanes. Actually, highway 401 in Toronto has 17 lanes in 6 separate carriageways. All these add width, which is converted to length! Such statistics may be useless for your driving but they're exactly what trans-portation authorities – and contractors – care when it comes to resurfacing the 'highway from Montreal to Toronto'. Effectively, this is an area estimate. Behind the 'lane' unit, there are more measurements. Lane widths range from 2.7 to 4.6 m (9–15 ft). The interstate system in the US uses 12 ft (3.7 m) standard width. Lesser widths may be found in lower classification roads, such as regional and rural. These all matter, even before additional consider-ations like medians, islands, shoulders, and more.

Then again, a commuter bus company between the two cities doesn't care much about any of that. Certainly they don't care as much as they care about *passenger-kilometres*, the number of passengers they carry over some kilometric distance. So do airlines for their flights. Transport companies care about *tonne-kilometres*. All these metrics indicate performance and are compared against capacity. Still, for finance officers in such companies the relevant unit will be *revenue per passenger-kilometre*. For those who care about congestion, traffic statistics per kilometre will be needed. If safety is the issue, additional statistics will be necessary. Knowing that night traffic accounts for 25% of all traffic on a stretch of road but for 55% of accidents has its own significance for potential corrective measures. There's always more than meets the eye. Definitions, methodologies, and other metadata will always be critical.

### Range, Not Point

There are more 'dents' to supposedly fixed physical measures. Some are fixed only under specific circumstances or certain conditions. The 'constant' $g$ for gravity assumes different values at different altitudes. Prototypes of the *standard* units were kept vacuum-sealed at a specific place as material expands and contracts at different temperatures. Moreover, the human desire for fixity is frustrated by the universe itself. The distances among celestial bodies offer an unlikely entry point to something that not only throws many people off-kilter but makes statisticians cringe too – through some combination of denial and self-pity (*"why do I have to drink this glass"* and *"I wish this weren't so"*). There are daily examples of people screaming at or begging a statistical guy or gal: *"I want the number, just give me the number – I need to know the distance to the moon, and I need it by close of day"*. He won't take no for an answer, but he won't take well to the true answer either – and he may be your prime minister's chief of staff. It turns out that the distance to the moon is not one number, it's many. When millennia ago Heraclitus declared '*ta panta rei*' (*everything flows*), he could have added *"and not in perfect circles but in ellipses and the like"*. This makes the distance to the moon not a constant but a range. The shortest distance (*perigee*) is 363,104 km and the longest (*apogee*) 405,696 km, for an average of 384,400 km. (Even these numbers smell of rounding rather than precision.)

The world is what it is, it's us who try to measure it. It'd be nice to get a single, accurate, and time-invariant distance to the moon but we have no grounds to complain why it doesn't behave to our liking. Economists know that blaming the economy for not behaving according to the assumptions of a stylized model is a step too far. People are starting to learn more and more about numbers and statistics. Some are reading this book!

The example above serves as a segue to the role of statistical producers in numeracy. For data originating in surveys, a range is what's really produced – we just discussed the existence of a margin of error, right? Still, this could come as a brick to the head of many users. The ostensibly unique estimates published historically were accompanied by quality indicators for this very reason – albeit in the fine print. Why the emphasis on point indicators? Well, most people couldn't take the range. They still can't. Such is the state of our numeracy. A cynic can always argue that masses need red meat and blame 'the media' (frequently unfairly) as they'll report a 48% as higher than a 47%

in a poll, when they're both subject to a 3% margin of error. Thornier issues arise, of course, when comparisons over time are required and the intervals for the estimates overlap. Keep this in mind for later, as this issue is in the 'cutting edge' of approaches to deal with data confidentiality.

The inability and discomfort to face a range estimate is a recognizable symptom of innumeracy. As we must come to terms with such limitations in our measurements, help arrives from an unlikely source. The weather, an area teeming with data, covering a whole host of variables and going back a very long time. Temperature data in particular are the stars of the show. They're applicable to specific places and they can range substantially throughout the day – in most places. Sometimes the range can exceed 20°C in a 24-hr period. Thanks to the weather channel, people have already been conditioned to think in terms of a range, presented as daily minimum and maximum temperatures. After all, we know that in our climate, temperatures will typically be higher in the afternoon than in the wee hours of the morning. We're now at the point to display temperature by the hour. Moving to a continuous clock is not far-fetched. Unlike the numbers we saw earlier, we can only put in the vault yesterday's weather data. They will not be revised. The forecasts will – but still that's one better than we can do with other statistics as we'll see later. To top it all off, any wrong prediction by meteorologists applies to the future, never the past.

From our discussion above, data accuracy should be understood as a useful idealization, to be adapted to the circumstances of the empirical uses we put them.

---

## Applied Cases

We've started to scratch the surface of some statistical matters. Simple applied examples can now help us ponder actual and potential *data sources* and their *intended use*. We can learn a lot from situations we can relate at a personal level before we tackle bigger societal and global statistical issues. Let's think together.

### Around the House

While still in the realm of physical measurements, we'll move from natural to man-made structures. These are closer to Euclidian geometry than nature, which has no perfect square, circle, or straight line for that matter. Along the way, we'll get help from our new character, Jim. Not one to burn all his spare time on the couch, he opts for a couple of do-it-yourself projects. And he really means to do a good job. Knowing full well the importance of measurements in such things, he even invests for the first time in a laser gadget for precision in measuring distance. Then, he's off to the first project.

Consider the measurement of the square footage of a bedroom for the purchase of flooring. Easy-peasy. Jim measures with his new gadget – twice. (To check the new gadget, he also went on his knees with the old measuring tape.) He transfers everything to a spreadsheet because the room had a couple of irregular corner niches. Double-checking his data entry and the calculations, the 178.8 ft$^2$ he arrived at is a darn good estimate. Jim is off to the hardware store. There, he'll heed the advice of the salesperson – and his own judgement – and buy a bit extra. Depending on the material and the packaging, that could easily be 10%–20% more. With the job well done, he's emboldened for the next project. But he can't shake the thought that he was too picky measuring – doesn't square with the left-over flooring material. He wouldn't repeat the mistake of overinvesting in such precise 'data collection' or 'data processing'. And he'll definitely ignore that corner niche with an inch here and there.

Jim moves on to painting (another room – first you paint, then you replace the floor as he found out). Even though a big window and a couple of doors made it trickier to measure, he surprised everyone. He was all done in mere seconds. He didn't fuss over the corners, didn't even use a ladder. No measuring twice, no spreadsheet. He was pleased with the 285 ft$^2$ he came up with, which he rounded to 300 ft$^2$. With no hesitation whatsoever, Jim buys two gallons of paint for two coats (at 3.78 litres, each gallon covers 400 ft$^2$). Surely, with very exact measurements he might have gotten away with a gallon and two litres, but not only he'd be taking a chance, it would cost him the same if not more. (That's how the business works, 'litres' are around 900 mL each and sold separately and much more expensive than a litre in a gallon). He feels he did well, the left-over paint is not worse than the left-over flooring. Perhaps he'll ballpark the third room with his eye.

There is a sense of irony in being able to measure accurately when it's not really needed. Accuracy is a desired quality, of course, but must be balanced against the need. The same goes for the amount of data needed but sometimes research enthusiasm (another quality) gets the better of us. *"What are you gonna do with the data?"* is a valid question upfront. I've repeatedly seen examples of overreaching, as if the fate of the entire world depended on one more survey question. But good measurements are metadata driven, that is, they're guided all along by the end outputs and their intended uses.

## In the Backyard

Let's spice it up a notch. Jim lives in a fairly new subdivision, on a street with about 50 houses. They all share the same straight-line fence at the back. As luck had it, something went wrong with the grass from the get-go, never quite caught. By now all backyards are ruined, and this has become the topic of conversation in the neighbourhood. After intense deliberations in community meetings, homeowners decide to pull together and negotiate a good deal to re-sod the whole thing. The first task is to come up with a very good estimate of the total size of their backyards. This will surely help their bargaining power. In a moment of collective enthusiasm, they name Jim to lead this. Jim is good at such things. (But we can step in and help along the way.)

Surely, the task is manageable. Jim only needs to produce one number. Not like a whole new data set that would then trigger data analysis and the like. Being the inquisitive sort, as soon as he gets home Jim starts thinking. *"How exactly am I gonna do it?"* I don't know Jim that well and I'm not sure how his thinking will progress but a likely first thought could be: *"I'll measure the length of the entire back fence and the distance from the back wall of my house to the fence. All I need is someone to help me with the measuring tape and I'll be done in an hour."* No big deal!

As he sips his coffee, things start to come to his mind – uninvited. Would that estimate be good? He just realized that his house is set farther back on the lot than his neighbour's. The length to the fence is much shorter, easily 4–5 ft. This could lead him astray to a bad estimate. *"Perhaps, I'll measure the distance to the fence for half-a-dozen different houses and take the average. That should do it. It'll only take an extra hour or so"* passes through his mind – but he doesn't quite let himself believe it. Come to think of it, some models don't have straight walls at the back like his, they have irregular shapes and recesses with different distances from the fence. Will he need a rodometer rather than a tape? But Jim won't let anxiety set in. *"I'll take 2–3 measurements for each house, take the average, and then use it to average across the half-a-dozen houses. It'll just take a bit more time".* His mind still doesn't let go. For one, he really starts not to be comfortable with that business of the average of the average... that would surely throw off his measurement. Would a simple average of the 2–3 distances be adequate? Some parts are wide and some recesses are very small (as in Figure 2.1). A weighted average would be more accurate. But then he'd need to measure the widths of those parts too, not only their distances to the fence. He'll probably have to make detailed drawings and do

**FIGURE 2.1**
Backyards

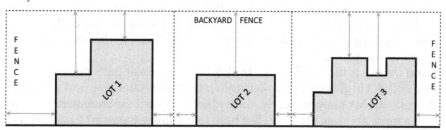

more calculations. Oh boy, it'll take more time. Worse, he could still end up with shaky estimates – unless there was a better way.

Perhaps. But first matters are bound to turn worse. As he mulls things over, and images of himself in backyards with tape-in-hand run through his mind, the lethal blow strikes. How on earth did he miss this! Most backyards have sheds, patios, and flower beds. A couple have swimming pools. They're all different, he can't possibly average over that too. He'll surely get in trouble. The thought that he may have to measure each and every backyard creeps scarily in his mind. Quick mental math tell him that even if he avoided chit-chats with the neighbours, it would easily take half an hour to measure each backyard. Not to mention the drawings, the spreadsheets, and the calculations. He can only devote a couple of hours a day at most, and there are days that he can't at all. This thing could easily drag out for a month. Is there no escape? What choice does he have? Since he took the job, he'll do it well. He'll be more careful next time it comes to volunteering. Anyway, he doesn't need to decide right now. He'll sleep over it. But let's just say, Jim didn't have a peaceful night.

The possibility of measuring all backyards in detail brought him more cold sweat, beyond that related to the tape work, the help needed, and the unavoidable chit-chats. "*If I am to measure each backyard, how exactly will I do it? I don't know everyone well. Do I just jump over from the back? Do I knock at the door, explain, and ask for permission? What if they're not there? How many times do I go? Do I leave a note to call me if they're not home? Should my wife know to answer in my absence? What if someone says no? Should I have a plan B?*" He can't cut to the chase. Perhaps, just a quick-and-dirty measurement with some rough allowance for landscaping sounds good again. Jim is not neurotic but responsible.

Within a few minutes, and all by himself, Jim suffered the conscientious researcher's curse – second-guessing, leave no stone unturned, leading to indecision and then agony. He thought of the pros and cons of alternative approaches, from simple to more involved, from sampling to a census. He identified trade-offs in effort, time costs, and accuracy of measurements. He even considered many of the logistics of a survey, how to deal with consent and permissions, non-response, backups for communications and the like. Far from his confident self, Jim turned unusually indecisive. He knows he doesn't need a perfect number but he wants a good estimate. The sod company charges by the square foot after all. But he won't despair, he just needs to cut through all this. (Another feature of most projects at this stage, before someone makes decisions!) Tomorrow he'll talk to that statistical guy he knows and trusts.

And *the guy* is the helpful sort. In short notice, he's at Jim's place for afternoon coffee. He gets the rundown, understands Jim's thinking and dilemma, and rolls up his sleeves. Surely, individual backyard measurements would yield a more accurate estimate. But how to minimize the pain? "*Your job is to* come up with a good estimate Jim but that doesn't mean you have to do all the work. Since this will help everyone, have them pitch in, spread it out. Ask them to measure their own backyard and give you the number", the guy says.

Part of Jim wants to breathe a sigh of relief. If only it weren't for the other part that realizes he'd relinquish control. *His* estimate would be as good as what the neighbours give him – and, he knows, some are not as meticulous as he is. But something has to give, right? Sensing his ambivalence, *the guy* offers refinements. *"You can prepare a few instructions and an example of how it should be done. You can even offer to do it for someone who can't. Give them a week and even if you only get half back, you'll have enough data to estimate the rest quite well."*

Jim starts to visualize dropping packages at the neighbours' doors and waiting for calls and emails. But snappily *the guy* hits again. *"You know…one way to eliminate measurements altogether could be property plans, all the dimensions are there. Homeowners have them or they can get them from the city"*.

Jim is now into admin data! Is there a rabbit in this bag or just more trouble? For a few split seconds his mind wanders. Can he really ask the neighbours, will they give him their plans? If he could get them from the city for all, does he need consent? Does he have to pay? How long will that take? But the guy is not done. *"Except for special circumstances, residential building permits are public records and can be put online. I believe the city publishes building polygons as open data. If all fails, you can apply under access to information – but that could be long and painful."*

Not exactly the cut-through Jim was looking for. This is all interesting at some level, he can see benefits if something big with thousands of backyards was involved. But can he afford that kind of hustle for what he's up to? *"Decisions, decisions, decisions"*, he utters. Knowing when it's time for a change in scenery, he also decides they've had enough coffee. *"Do you want a real drink?"* he offers.

With a clearer mind now, out come the big cannons. *"You know, I changed my mind"*, says *the guy*. *"There are always discrepancies between plans and actual construction. Then, they should have the swimming pools since people need a permit but it's doubtful they'll show the sheds, the patios, and all that. And, who knows, you may find out you're a full foot into your neighbour's property, not good…You won't avoid measurement!"*

There goes that idea. We don't want collateral damage now, do we? But Jim feels lighter. No rabbit but no applications and paperwork either.

*"Something just hit me that we're trying at work for some other stuff"*, the guy continues. *"Not sure it's for you but just to share…We started getting satellite data for geospatial statistics. The images are up to date and at high resolution, they show everything"*.

Jim is now a bit dumbfounded. He regrets offering that drink. How did satellites get in the mix now? *"And then what?"* he asks. *"Well, then someone can use algorithms and build an AI application to measure the green space in the backyards. Only the green space. If done well, it can be more accurate than measuring."*

It takes a few seconds for that to seep in. Jim will now react any which way you want him to react. But with his button pushed and his momentum unleashed, my version is: *"If it can be done with satellite images, why not with Google maps? I've seen my house and the whole neighborhood there. Everything shows. It's free and it's there now"*, he says. *"And why do we need any of that? What if I go around with my cellphone and take some good shots?"*

For a moment I thought he was going to mention a drone. *"Sooo? Can that other guy do it? How long will it take? How much will it cost?"*

*"That I don't know. We just started this project and put a whiz kid on it. We'll see..."* retorts *the guy.*

We'll leave the story here. You can mull things over in your mind, your way. I'll briefly take stock of some key moments.

In the course of a few hours, Jim took incoming from an avalanche of things flying his way, he rode a roller coaster through hopeful, too-good-to-be-true, and good-for-other-things ups and downs. From a no-big-deal-just-an-hour plan, he flirted with sampling and survey procedures. From total control in his bailiwick, he came face to face with the possibility of staking his reputation on the work of others. He was exposed to the vagaries related to acquiring and repurposing admin data, even open data, touched on issues of privacy, confidentiality, and access to information, and found out that soon measurements can come not from his tape but from pictures fed into some code. And to think that all he wanted was a good estimate for the grass in a few backyards. He feels grateful he's not in the big league, having to tackle the big things of our days. Who knows what else is out there to torture his mind.

---

## That Matter of Time

I can't talk about measurements and miss *time*. Our world is not quite complete with its three dimensions. Time is always lurking around. Brilliant minds told us of the space-time continuum. Whether time is a fourth dimension or it bends, I'll leave it to science. My interest here is numbers. We all have an appreciation of what time is and what it does. It's present everywhere and at all times (*sic*), marking irreversible occurrences. Of course, from the get-go humans felt compelled to measure it. In reality, we can only 'measure' the passage of time. Then, we use such measures to tell age, estimate periods, and all the rest.

Time flows constantly. Anyone would be excused to perceive it as a straight line. How come, then, we ended up with time measures that repeat themselves cyclically? It's a safe bet that the main culprit was the endless cycle between *day* and *night*, produced by the rotation of the earth on its own axis. Sunrises and sunsets kept coming. Much longer cycles were later identified as *seasons*, thanks to the rotations of the earth around the sun, eventually marking the *years*. We didn't really have to divide the year in the periods we now know as *months*, which derived from the *moon* whose very visible phases caused by its rotations around the earth also impressed us – enough to squeeze it in although it doesn't fit that well (the year has 365.25 days and the lunar month about 29.5 days, which leads to a *blue moon* every 2–3 years). We definitely didn't have to divide the month in *weeks* and each week in seven *days*. And there's nothing really behind dividing the day in 24 *hours*.

All the ancients took a stub at keeping track of time. The Sumerians, the Egyptians, the Greeks, the Mayans, and others created subdivisions of time and their own calendars. Somehow Julius Caesar gets credit for putting the Roman world on a solar calendar, and giving us the *leap year*. It was replaced late in the 16th century by the Gregorian calendar to correct inaccuracies that had accumulated. (Ten days just vanished; October 4 was followed by October 15, 1582.) Slowly, it was adopted widely but still we have no universal standard. The history of how we got what we know and use today is interesting. This includes naming conventions, such as how days and months were named after planets or numbers – in many Latin languages, September to December at some point stood for 7th–10th. (In modern Greek, the ordinal numbers second, third, fourth, and fifth are still used for Monday–Thursday – and in female form.)

For much of history, only the day (as in *daylight*, between sunrise and sunset) was divided into segments, sometimes equivalent to two hours, other times in 12 periods but not of equal duration. Daily life was largely governed by the sun, with *noon* as the benchmark. Such practices continued well into the 13–14th centuries, until the arrival of the early mechanical clocks that introduced a 24-hr day. (Still, the wide adoption of the 24-hr clock among countries had to wait until the 20th century!) The influence of the Sumerian base 60 prevailed among astronomers then, dividing the hour in 60 min and, eventually, the minute in 60 s.[‡] As we saw earlier, a metric system for time appeared at some point but it didn't catch despite the efforts of the French revolution. Thus, our existing measurement of time complicates calculation (just try to subtract 2:45 pm from 3:30 pm – on paper, not mentally). Interestingly, when we became able to measure at finer levels, we gave the *second* metric subdivisions. We use tenths and hundredths of a second, and sometimes the latter is still not enough to separate athletes in competitive 100-m dash races.

How we measure time is another example of a lengthy human invention. Predictably, we invented devices both to tell time (as in *what time it is*) and

---

[‡] The second is now defined as the duration of 9,192,631,770 periods of the radiation corresponding to the transition between the two hyperfine levels of the ground state of the caesium 133 atom.

to measure its duration (how long), which we use to calculate *age*. From the hourglass to the sundial, the pendulum and the mechanical clock, to today's digital and atomic clocks capable of unimaginable precision, the history of timepieces is fascinating and continues to this day.[§] We've even invented methods, based on a radioactive isotope of carbon, to date fossilized organic materials tens of thousands of years old. Then, we invented the meridian, the Coordinated Universal Time (UTC), and divided the earth in time zones. We even devised daylight savings for energy. There are volumes of information and you can read all about it. I will only share a couple of thoughts with you.

The first deals with the rather counterintuitive choice to embed so much circularity into the measurement of time. Any fleeting *moment* is unique, when it passes it belongs to the past, and there's always another moment ahead. A linear representation of time would assign unique numbers to each unique moment in history. There would be no years and no return to some Monday in December at 10 am. Such thinking has given rise to alternative approaches for the measurement of time, and different clocks. *Epoch time* (or *Unix timestamp*), used in computing, encodes a point in time as a number, equal to the seconds that have elapsed since midnight on January 1, 1970. *Internet time* was devised by Swatch in the late 1990s. Hyped as a fitting response to the digital era, it divides the day in 1,000 units and shows the same time everywhere – no time zones and, therefore, no need for UTC or 'your time or mine'? in international communications. It didn't catch. Star Trek's generous inheritance includes *stardates*. Time keeps moving ahead. Of course, as you can see in Figure 2.2, moving from spiral (or coil) to straight-line time would ruin people's anniversaries and birthdays. But then, rather than sharing birthdays we'd have our own, unique birthdates. I like the idea of unique moments moving forward but, to be fair, our system today is fully capable of pinpointing time with the use of seconds, minutes, hours, days, months, and years. Still July 1, 2020, at 12:00 am loses out to stardate 98097.68.[5]

The second thought relates to difficulties I personally encountered trying to teach time to kids. It came after we were mostly done with numbers and counting, knowing odd from even, parts from whole, organizing numbers in sets, even performing basic calculations. I thought it'd be a breeze. It was anything but...Beyond a totally mechanical understanding of our day and night, weekdays and weekends, that 1 pm comes after 12 pm and the like, it wasn't easy for the kids to grasp the essence of telling time or measuring its passage. How many hours had passed from lunch or how many minutes since we started were causing stumbles. And just when I thought some things had sunk in, I had to answer *"why is the 6 used for 30 minutes?"* Except for mix-ups attributed to the Sumerians, I had to answer for the different number of

---

[§]   To promote long-term thinking, an initiative underway is building the *Clock of the Long Now*, a
      mechanical clock that can go on ticking with minimum human intervention for 10,000 years.
      It started at year 01996, using five digits to avoid the Y10K problem.

**FIGURE 2.2**

Linear vs. circular time

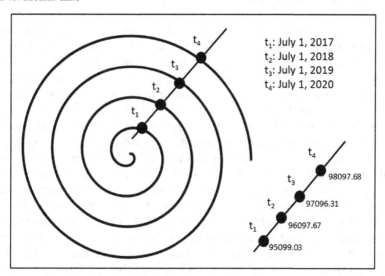

days in each month, defend February 29, explain why the number of days in a week are seven, and other such.

But there was another issue too due to an error I was committing, completely unaware. I was working with 5-min increments. Pointing to the long hand of the clock, I'd refer to 5:10 and 5:20. That's what I knew my whole life. It took a while before I realized what was happening. The kids were frequently looking at the digital clock on the television set. They had mentioned a time like 6:37 before but, being fixated on the hands of the clock, I didn't think much of it. When, unsuspectedly, a kid mentioned again 7:23, I clicked. I was trying to teach them to tell time in a way that was far less accurate than they could see – and already knew. The kids might not have been able to put 7 pm in perspective but they could read the TV clock, and they knew that was the time. I still have no childhood (or adult) memories with reference to 8:17 am and have never ever made appointments for 11:08 am. But we do live in a digital era, and one of higher accuracy. Talking about learning from kids…We have to let go of some ingrained stereotypes. Remember this for your next date.

## *Fact-Checking Tips*

In reality, there's no such thing as a straight line. There are always deviations between blueprints and actual construction. If not, at least one of the walls in your house would be perfectly straight. But we build skyscrapers with such blueprints. We build aircraft and go to space.

*An accident:* On December 11, 1998, NASA launched in space a robotic probe, the Mars Climate Orbiter, to study the planet's climate and atmosphere. Upon arrival, it was supposed to orbit the planet at an altitude of 226 km. To underscore the importance of measurement precision in such matters, this represented a tiny proportion of the distance from earth to Mars (0.000411%– 0.000057%), whose (theoretical) minimum is estimated at 54.6 million kilometres and its maximum at more than 400 million kilometres (another range hard to pin down, partly because it's a 'function of time')![6] As the probe was approaching Mars, a trajectory correction manoeuvre was computed and then executed on September 15, 1999. Subsequent measurements based on navigational observations, however, revealed that the probe's altitude may be only 150–170 km. Similar calculations the day before its orbital insertion placed it as low as 110 km from the surface. Despite the discrepancies between instrument-based and observation-based calculations, NASA didn't perform another trajectory correction manoeuvre. Alas, the promise behind the spacecraft would not be fulfilled. On September 23, 1999, it burned up and broke into pieces. Not a happy ending. Why?

Post-failure calculations showed that, far from an altitude of 226 km, the probe might have been on a trajectory that would bring it as close as 57 km to Mars, when 80 km was thought to be the limit for survival. The primary cause of the discrepancy in the orbital measures was that a piece of ground software supplied by Lockheed Martin, the manufacturer, produced thruster impulse data in *US customary units* while a second system that used these data as inputs for trajectory calculations followed NASA's conventions specs. In short, commands sent from earth were in imperial units (pound-seconds) without conversion to metric units (Newton-seconds). The cost of that mission was US$327.6 million.[7]

The morale is, of course, twofold. First, errors can happen anywhere – even when the stakes are as high as in NASA. Second, there's more to measurements than accuracy alone – all input data were accurate. You'll come across plenty of comparisons involving both lack of conversion and conversion errors. Exercise due diligence when you see todays' exchange rates used to convert data from 1990. Gain some familiarity with frequently used exchange rates based on the notion of *purchasing power parity*. It's used extensively, and economists consider it a (theoretically) superior way to compare across currencies – but it won't be of any practical use when you travel. There's easy access to all kinds of online converters, particularly for physical measurements.

*Back to the basics:* Fluency with the basics of numbers is essential. All kinds of errors occur from mishandling the basics before any higher 'sophistication' comes to play. Different types of numbers, their properties, units, scales, and limitations must be front-and-centre.

The old Greek *thermoscope* wasn't replaced by a single modern global standard to measure temperatures. The Celsius scale assigns 0°C to the freezing point of water and 100°C to its boiling point. Fahrenheit's equivalents are 32°F and 212°F. The Kelvin scale, used only in physical sciences even though it's one of the international standards, has its lowest point at *absolute zero*. At −273.16°C, molecule motion and kinetic energy are not possible. No matter the scale, assigned zeros are arbitrary. The implication is that 20°C isn't 'twice as hot' as 10°C. *Categorical, ordinal,* and *interval* numbers have no true zero. This is also the case for the variety of scales that measure seismic activity. The most well known is the Richter scale, which is logarithmic (base 10). The magnitude of a 7 Richter earthquake on a seismographer is ten times that of a 6 and 100 times that of a 5.[**]

Time-invariant measures, like as-the-crow-flies physical distances, are of limited applicability in socio-economic research. However, measures with different timing and frequencies are common. *Flows* don't have a quantity independent of the time period to which they apply, *stocks* are good for a point in time. The two-lane equivalent length of a highway network, a stock, doesn't change overnight – but is not fixed. There's no reason not to tell apart stocks from flows, when up-to-date monthly data on traffic are applied to severely outdated stocks. Frequency and timeliness issues affect the inter-pretation of data.

Many of the data errors you'll encounter can be traced back to the basics of numbers. Whether malicious plays or honest mistakes, they're a source of frustration. This is definitely an area where significant mileage on numeracy can be had without a huge amount of effort.

*BMI and mathiness:* Imagine someone invited in a medical conference to offer her expert statistical opinion on the BMI, the body mass index. (Imagining is like a dream, it may feel long but really lasts a few fleeting seconds.) The formula $BMI = weight/height^2$ figures prominently on the screen. Values below 18.5 indicate that someone is underweight, 18.5–25 normal, 25–30 over-weight, and over 30 obese.

At the first glance, she palpitates. Can this really be true? Weight will measure mass (muscle, fat, and bones) – fine. But she can't help looking at the exponent, intently. It acts as a magnet for her eyes. Hmm, there must be a good reason. *"I'm sure there must be a reason"*, she reassures herself. This is followed by *"I hope there's a good reason"* and *"There better be a good reason"*. As if to save her time researching the matter, the presenter explains: *"Physiologically speaking, people have different frames and bone structures. Also, weight increases proportionally much more than height, and we want the BMI to remain the same when all body dimensions double. A correction is needed"*.

---

[**] This is so when earthquake amplitude is concerned. For the amount of energy released the difference between units is 31.6 times.

She knows the mathematical square-cube law. If the size of a cube doubles from 1 cm to 2 cm, its surface area will increase by a factor of 4 ($2^2$), and its volume by a factor of 8 ($2^3$). That's why gigantic beings are a myth, they'd collapse under their own weight.[8] Short of helping her, what this explanation really proved is that the BMI formula would give the erroneous impression that very tall people are fatter than they really are, and very short people thinner. Sure enough, as if he read her thoughts, the next presenter proposes the Ponderal index, which allegedly compensates for that by using 3 as an exponent (BMI = mass/height$^3$). She doesn't see *how*. The last presenter argues that the exponent 2 is too large for short people and too small for tall people. He offered a compromise on the exponent and added another factor: BMI= 1.3 × (mass/height$^{2.5}$).

Whichever way one goes seems to muddle the matter further. She feels that this is not the right course to follow. It bears resemblance to what Carl Bergstrom and Jevin West in *Calling Bullshit* called *mathiness*, dressing something that makes very little sense with the veneer of a scientifically looking equation. It's not as scary as H = (F, SL, M$^3$)/bJ, where H stands for *happiness*, F for *family*, SL for *social life*, M for *money*, and bJ for a *bad job*. But what's really this whole BMI thing all about? The basic question is whether someone with a certain height and build is overweight or not. Why not deal with this straight up? The fact that people of equal height can have different frames or bone weights (measured at the wrist?) is fine. But for each individual all these stay constant. So, she offers: "*The choice of the exponent is shaky. But whatever exponent you use, the denominator will always have a fixed value. Where's the sense in having a formula with two variables, only one of which changes? I think you can safely just monitor the weight of someone with a given height and body structure to see if he or she is overweight. That's all that matters – certainly until they start to shrink late in life.*"

*Red flags:* After a lifetime in the hospitality business, a new retiree is intent to follow his true passion. He now has the time to bring out the amateur astronomer in him. He's always been attracted to science, he loves observations and measurements. That's how you separate fact from fiction, truth from falsehood. Diligently, he buys his dream telescope and other supplies. Searching online, though, one thing leads to the next. So, he also invested in a gravitometer he happened to come across. It'd be great to reproduce gravity measures. Who can argue with reproducibility? This is key to all science. Upon arrival, he sets out to a nice hilltop to measure gravity. To be sure, gravity is tricky. Because the earth is not a perfect sphere, it's slightly higher in the poles than the equator. It's also affected by density and altitude, it's lower at sea level than high up a mountain. But we're talking small differences, it never ventures far from 9.8 m/s$^2$. He knows all that anyway. So, he's taken aback when his first measurement is not in the ballpark. His instrument shows 11.234567 m/s$^2$. Strange, he thinks. With persistence and determination, he takes more measurements in different locations all with similar results. Thoroughly puzzled, having no business with peer reviews and the like, we live in a free society and all, he blogs about his findings. In good prose, he shares his experience asking how this can be possible. Some of his writing could have been a touch clearer. *"Never mind how many times and in how many places I measured, I got the same result for gravity – 11.234567m/s$^2$ and not 9.8m/s$^2$"* could be construed by someone as an innuendo, questioning the *g* constant. What happens now?

Most people will never even know that he has a blog. A few may come across it, giggle, and ignore. But one – we know – will step on it. He will share. And, he'll add *his* stuff.

*"You see? More evidence. They're lying to us. It's not only gravity, you can't believe anything. I told you it was a made-up video, nobody really went to the moon."* A whole discussion will ensue, among prone circles. *"You're right, I agree. We have to tell the others…and, by the way, the earth is flat. I've been driving for hours and it's all straight."*

Whether this goes viral or not, at some point someone will call him out. He'll heed the call, and some back-and-forth will settle the matter. Being who he is, he'll issue a statement making it clear that he never meant to question the *g* constant. He knows better. He's not a cheat or a conspiracy theorist. It'll turn out that the gravitometer he bought wasn't much of an instrument, it was really a toy. (Actually, the real instrument is called *gravimeter*.) However, the rest of the chatter will have by now a life of its own. The 'pressures' to silence him and bury his 'discovery' will also be added.

I made up this story, but many more with the same basic characteristics are in our midst. Some may be more nuanced. Why should we have absolute faith in all the cosmological standards? Isn't science all about questioning and reproducing results? Why should we believe the government's CPI? What if I produce my own gross domestic product (GDP), blog it, and find myself with quite a following? Tricky stuff. There's also a well-documented bias against publicizing results that confirm existing wisdom. It's unlikely that the fellow in our story would have blogged to confirm the *g* constant. It's doubtful that I'd get my 5 min of fame if my GDP didn't go against that of the statistical office. Such is our state of affairs. By the end of this book, you'll know that in our times the world of data is pretty much on the offensive. However, there's no choice but to invest in defence too. There's a need to fight back, some things can't be left hanging out there. Look out.

## Notes

1 Practically, we only use few. In the metre, for instance, we go as low as the millimetre. But science needs both cosmic and subatomic accuracy. Examples of such

measures are the gigametre ($10^9$), the zettametre ($10^{21}$), the nanometre ($10^{-9}$), and the picometre ($10^{-24}$). Many more exist. See, for instance, Wikipedia contributors, "Metre," *Wikipedia, The Free Encyclopedia,* https://en.wikipedia.org/w/index.php?title=Metre&oldid=1082887742

2    After improvements in 2011 in the caesium fountain clock operated by the British, the atomic clock allegedly won't gain or lose a second in 138 million years! Not to be outdone, in 2014, the Americans launched one that won't lose a second in 300 million years and made another version for the Italians in 2016 that does one better. Wikipedia, "Atomic Clock," *Wikipedia,* https://en.m.wikipedia.org/wiki/Atomic_clock

3    For instance, in 1781, the metre was defined as $^1/_{10,000,000}$ of the quadrant of the earth's circumference running from the north pole through Paris to the equator. Expeditions were organized to measure the relevant arc of the earth. A century later the international prototype metre was *"the distance between two lines on a standard bar of 90 percent platinum and 10 percent iridium"*. With modern advances, scientists in the 20th century wanted units independent of any physical artefact. So, in 1960 the metre became *"equal to 1,650,763.73 wavelengths of the orange-red line in the spectrum of the krypton-86 atom in a vacuum"*. With more advances in laser measurement techniques, values for the speed of light in a vacuum became astonishingly accurate. Since 1983 (with a few tweaks after) the metre is defined as *"the length of the path travelled by light in vacuum during a time interval with a duration $^1/_{299,792,458}$ of a second"*. Bureau International de Poids et Mesures (BIPM), "SI base unit: metre (m)," *Bureau International de Poids et Mesures,* www.bipm.org/en/si-base-units/metre

4    David Berry. "Trans-Canada Highway." *The Canadian Encyclopedia.* Historica Canada. Article published February 7, 2006; Last Edited January 4, 2021, www.thecanadianencyclopedia.ca/en/article/trans-canada-highway

5    Check the converter at STO Academy, "STO Stardate Calculator," *STO Academy,* www.stoacademy.com/tools/stardate.php

6    See TheSkyLive, "How Far is Mars from Earth?" *The Sky Live,* https://theskylive.com/how-far-is-mars#:~:text=The%20distance%20of%20Mars%20from,as%20a%20function%20of%20time

7    NASA, "Mars Climate Orbiter," *NASA Jet Propulsion Laboratory,* www.jpl.nasa.gov/missions/mars-climate-orbiter.

NASA, "Mars Climate Orbiter," *NASA Science, Solar System Exploration,* https://solarsystem.nasa.gov/missions/mars-climate-orbiter/in-depth/. Everyday Astronaut, "Metric vs Imperial Units: How NASA Lost a $327 Million Dollar Mission to Mars," *Everyday Astronaut,* https://everydayastronaut.com/mars-climate-orbiter/. SIMSCALE, "When NASA Lost a Spacecraft Due to a Metric Math Mistake," *SIMSCALE Blog,* www.simscale.com/blog/2017/12/nasa-mars-climate-orbiter-metric/

8    A 24-ft giant, four times the size of a 6 ft person, would collapse under his weight at nearly 10,000 pounds (4,500 kg). In this example, the scaling factor for the weight is 64 ($4^3$). The estimate comes from Science World, "What if humans were giants?" www.scienceworld.ca/stories/what-if-humans-were-giants/, November 19, 2016. Now, humans are three dimensional but not cubes, and weight doesn't need to change the same as volume.

# 3

## Humanity's Numbers

We dealt with numbers and counting in Chapter 1. We discussed measurements, units, and instruments in Chapter 2. This chapter creates a bridge from counting to estimation, before we move to the quantification of abstract constructs in the socio-economic realm.

### Just Imagine...

A wandering alien salesman makes a pit stop on earth. He asks the first earthling he encounters: *"How many of you are here, and is your number growing"*? Straight for the jugular. How big is this market and what's the rate of growth! In response he gets: *"Not sure, there's too many of us, billions I think"*. Not exactly what he was expecting...but it could have been much worse. *"Some people!"*, he thinks to himself – and the thought of throwing out an unflattering epithet crosses his mind. But he quickly decides to take the high road, conceal his sense of superiority (*"I'm the one with the spaceship after all"*), and up his game. He'll use his considerable perceptive abilities and pick a more intelligent dummy. Realizing also that his question was not the best for an advanced communicator, he refines it and moves on to the next target. *"How many of you are on this planet, what's the rate of change in the last 70 years, and how long do you live?"* he unloads. It so happens, she's a statistician with the United Nations Population Unit (UN) and his new answer is: *"As of July 1, 2020, there were 7,599,987,765 people on the planet, a new record in our history. Our annual compounded rate of growth between 1950 and 2020 has been 1.6%. The population of the Earth has doubled in the last 50 years, and we're currently adding one billion people every 12 years. Overall life expectancy is now 72.3 years, 74.7 for women and 69.9 for men"*. She also handed him a printout with the information (Figure 3.1). In your face alien!

"*Now you're talking*", he murmurs. From a non-answer to a detailed account, complete with history and forecasts. He's pleased with *his* choice of this earthling. He got more than he bargained for – except for that piece of paper, of course. He had come across on the *cosmonet* once that his own ancients too used printouts to display data in two dimensions before their stunning 3D holographic visualizations. He jotted in his notes: *"Not a bad*

DOI: 10.1201/9781003330806-3

**FIGURE 3.1**

The evolution of the human population

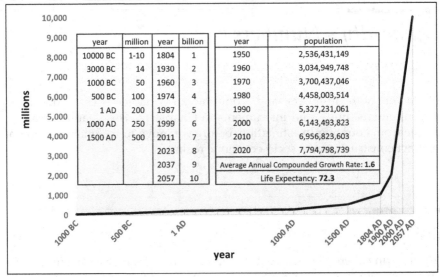

| year | million | year | billion |
|---|---|---|---|
| 10000 BC | 1-10 | 1804 | 1 |
| 3000 BC | 14 | 1930 | 2 |
| 1000 BC | 50 | 1960 | 3 |
| 500 BC | 100 | 1974 | 4 |
| 1 AD | 200 | 1987 | 5 |
| 1000 AD | 250 | 1999 | 6 |
| 1500 AD | 500 | 2011 | 7 |
| | | 2023 | 8 |
| | | 2037 | 9 |
| | | 2057 | 10 |

| year | population |
|---|---|
| 1950 | 2,536,431,149 |
| 1960 | 3,034,949,748 |
| 1970 | 3,700,437,046 |
| 1980 | 4,458,003,514 |
| 1990 | 5,327,231,061 |
| 2000 | 6,143,493,823 |
| 2010 | 6,956,823,603 |
| 2020 | 7,794,798,739 |
| Average Annual Compounded Growth Rate: **1.6** | |
| Life Expectancy: **72.3** | |

*size market. Future potential high. Very slow start but now they're really firing... they learned how to do it. If they stay on track, very soon we can sell them real estate someplace nice"*. He moved on.

We're here, though, and this is our story. Funny that despite his calibre the alien got the numbers and left. I'm still surprised he didn't ask for metadata. Probably not the best specimen of his race – which may be why they have him wandering in faraway corners. How good are these numbers, if any? Let's use this scenario as a springboard to delve into a few data matters. There's a lot to unpack.

## Count Me In

For now, let's just say that the population figures for the last seven decades are *good enough*, those for times long past are *purely conjectural*, and the numbers for the future are *as good as their underlying assumptions*. All this will be explained as we work through this chapter.

Where do the planet's population data come from? The UN is no more the source for these data than a lake is for the rivers that flow into it. While a lake is an amazing resource, the rivers need to be traced all the way back to their trickling beginnings, one at a time. World population statistics are a compilation of national estimates produced independently by about 240 countries and territories. When put together, the resulting dataset is as good as its sources. This type of data compilation involves much more than mere addition. It represents a key statistical function with applicability in many environments.

So, how do countries produce their population data and why are they not perfect? Why is *estimation* a more apt term than *counting*? Surely, we can count a group of living and breathing human beings accurately, particularly if confined in a given space, say, a classroom full of students, a congregation in a church, even all ticket holders in a sports arena. If need be, we can follow our ancestor and carve a notch for everyone on their way out of the door. 'All people' is another finite group, just larger. A country or even the planet are just bigger 'rooms', no one can escape our counting prowess. In principle, there's nothing wrong with this argument. In actuality, the classic gap between theory and practice is what transforms what should be counting to estimation. Let's dive right in.

## The Census

Censuses are the cornerstones of population estimates in most countries. Taking stock of entire populations feels conceptually clean. It boils down to enumerating a big group of people. Still, those involved even peripherally with a national census know that it's a big task. The exercise is fraught with a host of conceptual hurdles as to *whom, where, when*, and *how* to count. It's also logistically challenging. All kinds of questions pop up. To be sure, censuses are not merely statistical exercises. They get frequently caught in the geopolitics of our times. Our world is functioning on the principle of sovereign country-states after all. Sovereign countries are supposed to care about all their people, regardless of where they happen to be at a point in time. But people are mobile, and their lives can be eventful and complicated – some more than others. Where does a country end and a nation begin? Where exactly should the focus be? Inevitably, such questions spill into the census. The discussion below is meant to lead to some appreciation of issues.

*Whom to count?* Conceptually, censuses can align either with the notion of a *country* or a *nation*. In pure form, the former means that everyone within the geographical boundaries of a country during the designated time should be enumerated, regardless of status (*de jure*). The latter implies that only residents of the country should be enumerated, regardless of where they happen to be at the time of the census (*de facto*). Looking at a country as a well-defined geographic area can be well justified. But so is caring about a nation. Expats can attest to that. Decisions on the in-scope (target) population must be made.

Not surprisingly, different countries opt for different approaches. The *de jure* approach leads to the enumeration not only of citizens but also of immigrants, refugees, international students, workers on work visas, even vacationers, but it excludes citizens who are abroad at the time. Some may be away for years, some for a few short days, yet others may follow a periodic pattern of being in-and-out of the country. Such is the case of the Canadian *snowbirds* who spend routinely time in Florida, for example. Filipino expats around the world stimulate a good chunk of their country's economy. Many other countries care about their diasporas too. Can you miss cases like these in a census? Practical approaches by different countries attempt to strike a balance between the two concepts by using a hybrid approach, but their application is not uniform.[1]

*Where to count?* Censuses typically enumerate not only people but dwellings too. In fact, they start by finding houses before they search for people. (The first census in Uganda actually enumerated *huts*, not people.) The *whom* is linked to the *where*. A precise definition of residence is needed, not only to decide who should be included in the population count but also where they should be included. The distribution of the population within a country is needed for many and serious reasons. Most countries follow the notion of a *usual residence*, where people 'live and sleep' at the time of the census.* Yet, some people have no fixed address (homeless) and others have more than one, such as owners of multiple properties (such is the state of affairs) or are the children of separated parents with alternating living arrangements, perhaps in blended families. And what about people who live in *group residences*, including student dorms, asylums, and even prison inmates? Is it fair that they should inflate the counts in the municipalities where such facilities are located?

*When to count?* Historically, censuses have a designated day. Being very large undertakings, you can appreciate that *census day* is now more of a symbolic day, helpful to raise awareness in promotional campaigns. Collection continues for a long period after, frequently for weeks or months. Either way, the timing of the census can affect the count, particularly in conjunction with the choice of the in-scope population. (Can you imagine a *de jure* census day in a Mediterranean island in July when tourists vastly outnumber the locals?) In principle, many decisions are expected to be based on the usual residence "*on or around*" census day. This can create interesting circumstances. Newborns must be counted even if they first go home on census day and haven't slept there yet. Deaths on census day must also be counted, so long as the person "was alive for any length of time" during that day.

*How to count?* The traditional way of census taking was *door-to-door* canvassing. This followed earlier enumeration of houses, complete with sketches and drawings. Within a relatively short period, an army of enumerators would be hired, trained, and deployed in all corners of

---

* The origin of this dates back to the text of the 1787 US Constitution, which mandated a decennial census.

a country – on census day and for the necessary follow-ups. China still hires more enumerators than some countries have people. This practice continues but in many countries it's now meant to complement *self-enumeration* of families. This relies on mail-outs of forms and is now increasingly migrating online. Effectively, part of the census is outsourced to the people. A family member will fill the form on behalf of all the individuals in the residence.

## Clouds Ahoy

With all this noise, fuelled by the sheer number of people involved and the multitude of local decisions that need to be made, it would take a miracle for errors not to creep in. But no miracle here, errors of both omission and commission do occur. Dwellings and people who should have been included are missed. Others may be somehow double-counted. Even with well-defined and accurate in-scope population and residence, there will always be enough instances in which the criteria are not applied in a uniform and consistent way or different value judgements are exercised by enumerators and self-enumerated families. There are outright refusals too (but let's not get into the mandatory legal status of censuses and the threat of penalties). Rather than a civic duty, some may see the census as a chance to protest against big government intruding in their lives. Some may be afraid to go on record for immigration or other reasons. Others may have more important things to do.

Whatever the reason, design errors, honest mistakes, or non-compliance, there's always *undercoverage* and, to a smaller extent, *overcoverage* in a census. These can introduce biases in the results due to different characteristics between the populations involved in each. As one example, under-enumeration of the youngest age groups (under ten) is common. So, effectively, the census is treated like a massive survey, subject to response rates, imputations, and post-census evaluation through special methods (e.g., reverse record checks). In the case of the 2016 census in Canada, the *net under-count* was estimated at 2.4% of the target population. But net undercounts are also estimated for various breakdowns, such as province, gender, age, and many more. The estimated net under-coverage for *separated males* was a whopping 17.7%! Denial, denial, denial...

With the census as a benchmark, intercensal programs use *vital statistics* and *net migration* data to update population estimates annually, and in some countries sub-annually.

## More than a Count

A census is a large-scale undertaking. Ramping up the necessary infrastructure is labour-intensive, takes time, and costs money. The argument that too many resources are spent for one number (how many are we?) is not unheard of. But census-taking is not about arriving at a good population count once every five or ten years. If it were, at least nowadays, we'd probably have come up with some smarter alternative. Star Trek's *Enterprise* could take readings of all people alive on a planet below – not sure if it was based on heartbeats or something else. If it were a headcount we're after, I could imagine a campaign to get all Canadians out on the Trans-Canada (the 7,821 m long highway that runs from the Atlantic to the Pacific) on a Sunday in May, at 10 am and get the civic obligation over with. Whether holding a GPS device sent by the government to everyone, a satellite group photo, or something else, we'd be pretty much done – nice and early.

In case there are any misconceptions, a census is a lot more than a good population count. It collects detailed and coherent information on housing, family structure, education, incomes, and several demographic and socioeconomic characteristics of interest. Its content varies from place to place to meet the needs of individual countries, and over time to meet evolving needs within countries. Determining its content is always a balancing act between competing interests and availability of 'real estate' (reasonable questionnaire length). With decades of accumulating data, the census becomes the storyteller of a nation. Still, it's fair to bear in mind that censuses hail from the past, when other information sources were nowhere to be found.

The information made available by censuses is used virtually everywhere, oftentimes even unknowingly. It'd be a fool's errand to take a shot at an exhaustive list. Governments use it for the transfer of large sums of money inside a country. They use it for the planning of health, education, social security, public transportation services, and everything else. Businesses use it to plan their operations, expansion and location of new outlets, as well as promotional and marketing campaigns. Democracies use it for the redistricting of electoral ridings. Census data shed light on population movements and changes, record societal trends and evolution, and identify trouble spots of all kinds. They are indispensable to the research capacity of a country and internationally. Alone or combined with other sources, the uses of census data have literally no end in sight. It's simply incomprehensible to fathom the functioning of a modern society without such data.

Less visible to the public is that a good chunk of the statistical infrastructure inside statistical agencies is driven by the census, which exerts a dominant influence on investments in new methods and technologies. It serves as the universe (frame) from which random and stratified samples can be drawn for household or housing surveys, enabling the production of more data. It's frequently a unique information source for subpopulations of interest, including small areas such as neighbourhoods or rural areas for which administrative data are hard to come by and sample surveys can't produce reliable estimates.

Census data are disseminated in aggregate form to protect the privacy of individuals and are subject to strict confidentiality and security procedures.

This can be an irritant to inquisitive researchers but, except for very particular circumstances managed by detailed access protocols, we all have to wait for a very long time after the reference year for census microdata to be made public. For example, 76 years in the US and 92 years in Canada. For anyone interested in the whereabouts of others today, social media is a safer bet.

---

## The Tribulations of a Decent Synthesis

Back to global data and the UN. Or the US Census Bureau, which also compiles population statistics for all countries since the 1960s. Or any compiler of higher-order data for that matter. On offer are two crucial functions, aggregation and establishment of standards. They present an opportunity to look at how data are really produced, and trace their flows all the way to their sources.

### The Act of Aggregation

The first function is to synthesize the data, which is precious to us all. As already mentioned, this is not a matter of mere addition that can be accomplished through some mechanical operation. It involves painstaking work through a series of tasks that add significant value to the data. The basic steps of the process apply equally to an international body producing global aggregates from national data, a national institute producing country-level totals from subnational data from states, provinces, or other regions, even a unit within a private enterprise consolidating data across the operations of different establishments or divisions. The alternative, of course, is the centralized production of the desired data. Most times this is not possible because of jurisdictional or other issues. Even if it were, it might not be a good idea. There are supposed to be advantages in distributed production systems after all, right? But there's a price...

On top of all matters that affect the quality of national estimates, international aggregation conducive to cross-country comparisons has to put up with many more. Among them, I don't count the fact that the exercise includes about 240 countries or territories. It's a lot of work, for sure, but good people can handle it. So, what more is there?

- To begin with, a fair amount of back-and-forth with national authorities regarding their respective data is necessary. Our discussion so far makes it clear that there is no shortage of issues to deal with and clarify, from concepts chosen and methods employed, to quality assessments and supplementary sources, all the way to the nitty-gritty logistics of timelines and means of delivery. With sufficient understanding, many adjustments will be necessary as partial compensation for different standards or their non-uniform application. Things can become trickier quickly, in ways that could compromise the coherence of the data. Just as one example, if all countries conducted *de facto* censuses, adding up their numbers would lead to overcount due to duplication. Think of all those with dual nationality who'd be in the counts of at least two countries. A host of issues such as this may trigger more deliberations, the search for additional data sources, the need for subsequent adjustments and the like. You get the idea.

- Then, national censuses are conducted at different times across countries. Many take them every ten years, a few every five, others whenever they can. Some census years end in 1, some in 0, or another number. They also take place at different times of the year. The latest US census was conducted in 2020, and results were not available in time for the data behind Figure 3.1. Most European countries had their last censuses in 2021, Japan in 2017, Canada on May 11, 2021, and so on. So, how are the data referring to July 1, 2020 produced?

National estimates are benchmarked to the last census and projected forward through a variety of data sources, most notably registers of vital statistics. Then, to put everyone on equal footing, additional assumptions are made and temporal allocation techniques are employed. This means further interventions in the data by the compilers.[†]

- Next is the elephant in the room. Never mind how well the compilers perform all the previous tasks, their efforts are put in serious jeopardy by data gaps. When the major sources of global population estimates are national censuses, what happens when they don't exist? And for many countries they don't. *"One in five Africans is a Nigerian"* is a common saying in the continent – but sometimes it's one in six. We'd know which one is closer to the truth, had it not been for the fact that reliable figures don't exist – either for the numerator or the denominator. Nigeria, no doubt a populous country, has achieved notoriety for its inability to conduct a trusted census. A few other African countries have never had any. Many more don't have a current or particularly reliable one. The

---

[†] Statisticians compiling aggregate business statistics are familiar with the process known as *calendarization*, used to convert data from different fiscal years to the same calendar year. Some businesses choose fiscal years ending on June 30, some on March 31 (as does the government in Canada), or any other time. In the absence of calendarization, and depending on the circumstances, data for '2020' could include businesses with fiscal years ending at any time during calendar year 2020.

same is true elsewhere.‡ Population estimates for many countries are best-effort projections from any national and international sources that compilers can put their hands on. At times this is so because of wars and conflict, at other times due to political wrangling and the reluctance of some groups to face up to potentially inconvenient truths that could shift the balance of powers. Yet, there are times when it's due to under-resourced exercises. What impact exactly such gaps have on the overall estimates cannot be completely ascertained.

All the adjustments made and the caveats that remain need to be documented and become part of the *metadata* that will accompany the data. Looking at published tables with thematic compilations of statistics across countries, don't be surprised if occasionally you see more footnotes, endnotes, and asterisks than numbers. This may irritate someone looking for a quick fix but it serves the purpose of informing the users and, implicitly, defending the reputation of the source. But users of data have responsibilities too. If you don't see enough of that, ask for the metadata. In the case of world population statistics, they will come in the form of an encyclopaedia – one volume for every country, with a synthesis upfront. Imperfect as the results may be, this type of work is extremely valuable. Suffice it to say that in its absence individual users would have to recreate it. Only then we have access to a ready source, whether in search of a global aggregate, a subtotal for high-income countries, or any other grouping.

If you think that all this is more than enough to arrive at the numbers, from a user's perspective things can become much more interesting. The estimated counts do not stand still, they morph and multiply. Existing data are revised and new data are produced – which will also be revised. Updated infor-mation, between-censuses and sub-annual estimates, adjustments to facili-tate cross-sectional and temporal comparisons, all lead to a proliferation of numbers. This happens at both the national and international levels. Some countries even produce a different set of annual estimates with the July 1 date. And, of course, all numbers are spread wide on the internet where many get left stuck back in time. Looking for the population of a given country and year, you'll encounter multiple estimates from its statistical office, inter-national sources, and datasets in online junkyards. Even without any such extra complications, a researcher looking for the population of Canada from 2016 to 2020 will come across different quality vintages of estimates. One of the footnotes in a table from Statistics Canada is an example of its matter-of-factly and arresting transparency: *"Estimates are final intercensal up to 2015,*

---

‡ The Democratic Republic of the Congo, Eritrea, and Western Sahara have never had a census. Madagascar has not had one for almost 30 years, Somalia for more than that. War-torn Afghanistan and Iraq have not had a census in recent decades either. Less populous Lebanon hasn't had one since 1932.

*final postcensal from 2016 to 2019, updated postcensal for 2020 and preliminary postcensal for 2021"*.[2] You look, you find.

Much like every subject matter area, there's more than meets the eye in national and international population counts. Familiarity with the body of knowledge that exists, the metadata, could easily put a deep dent in any possible arrogance related to our ability for perfectly accurate counting of humans. As the US Census Bureau puts it *"… the current world population figure is necessarily a projection of past data based on assumed trends. As new data become available, assumptions and data are re-evaluated and past conclusions and current figures may be modified"*.[3] 'Current' population refers to times before 'today', of course. So, even yesterday's data are projections of sorts! The joke about meteorologists getting predictions wrong only when the future is concerned may sound less funny now. Such metadata didn't make it into the alien's notes. However, world population estimates are considered a good fit for most uses we put them to – just without the allure (or illusion) of extreme accuracy. In fact, for many individual countries they're quite good but this is moderated by gaps elsewhere. Rough estimates claim that nine in ten people are counted in a census today, up from about one in ten in 1850. Which is also why, on the whole, I told you that the numbers for recent times in Figure 3.1 are 'good enough'.

## Harmonization

The second function performed by data aggregators is of a longer-term nature. While it overlaps with aggregation it's more forward-looking. Whoever gets involved in such work comes to appreciate at no time the importance of common standards. When concepts and definitions don't align, collection and estimation methods differ, and reference times don't match, data synthesis becomes daunting. Regardless of how much effort is put in, it's messy. A harmonized approach up-front among all involved would lead to superior results, which leads to ongoing efforts for the establishment and adoption of common standards.

Harmonization is challenging even within the context of individual countries, but it's particularly difficult internationally. Our world has become global but we don't have a global government. Most of our affairs are driven by national imperatives, and statistics are no exception. At times, seemingly good standards may not fit well in the context of both developed and developing countries. Priorities and interests may diverge. Negotiations and compromises may be needed. In long-lasting processes, practical matters may also cause delays, such as frequently changing national delegations. Even in areas with long histories and successes, such as the UN's manual for national accounts or the IMF's manual for balance of payments statistics, standards and revisions are not all applied the same, if at all, or at the same time by all countries. At any given moment, there's a variety of applied approaches.

Training and capacity-building are frequently needed for the implementation of standards, even financial assistance. Such value-added work needs to continue. Ideally, in new subject matter areas, coordination of approaches and harmonization of methods and standards should precede actual measurements and manage to get ahead of the game when such opportunities arise. The OECD did a decent job when new measurements with no past were needed for the internet and the information society, with eventual help from the European Union's (EU's) legislative powers that standardized the work across member states and, later, from the UN that carried the message beyond.

No statistical development happens in a vacuum or is of the black-or-white variety. At times, statistical constructs and measurements respond to political imperatives, something that necessitates the balancing of competing viewpoints. The example of the Information and Communication Technologies (ICT) sector is instructive. Conceptualized as a vertical integration of manufacturing and service industries when the digital era was upon us, the unanimous sentiment that the sector should include manufacturers of ICT gear was not equally shared for ICT wholesalers and retailers. However, some multinational ICT players are manufacturers only in countries where they have facilities. In other countries, their substantial presence is classified under wholesale or computer services. You can understand why delegations from those countries would insist on a more inclusive definition of the industries that would constitute the sector. This is one example where statistics are called to support a constructed reality – something we'll explore more in subsequent chapters.

## Go Forth and Multiply

We'll now continue our assessment of the rest of the numbers given to the visiting alien. He certainly was quick to pocket them.

### Rewind to Blurry Shadows

Considering the predicament of our numbers today, what of our old numbers? Where did they come from? The oldest count resembling a census is supposed to have happened in ancient India around 330 BC. Early counting efforts have been recorded in ancient Egypt, Persia, China, and India, with several more in Greek city-states. The Old Testament contains references to the counting of Israelites after the Exodus, in conjunction with taxation – something that still instils fear. Census-like counting acquired a sinful undertone, with lingering effects, after a headcount by King David allegedly brought a plague as punishment. King Solomon even counted all foreigners – back then. The Romans too conducted censuses of sorts to keep track of adult males fit for military service (we said data must match a need). Much later, we encounter an extensive exercise in England that targeted land and live-stock more than people, ordered by William the Conqueror to ascertain the tax base. It culminated in the famous Domesday Book of 1086 AD, which gave rise to similar efforts elsewhere. Several more counts took place later, par-ticularly concentrated in cities. Jean Talon is credited with the first census of Canada in 1666 AD, documenting the need to bring more women to the new colonies. Famously, around the same period, William Petty's attempt to count the population of England and Ireland aspired to base policymaking on his method of *political arithmetic*.[4] Eventually, a decennial census was mandated in the US Constitution before censuses started to expand in several countries by the 19th century.

Beyond this synoptic account, there may be a bit more here and there but not much. Our numbers became good enough in the second half of the 20th cen-tury and have a degree of plausibility only over the last couple of centuries. Historical records for our distant past don't really exist. However, it's never a good bet to underestimate human curiosity and the lengths to which we can go to unearth things about ourselves. We set out to trace our origins, from hominids to homo sapiens, understand how we evolved, played with fire, what else did we do, including how many we were. Paleo-anthropologists, paleo-demographers, and many other brave souls have strived to provide guesses – and they're still at it.

The population guestimates of Figure 3.1 make maximum use of any kind of ecological, ethnographic, or archaeological proxy available, combined at times with dating techniques and even modern molecular biology.[5] The number of human settlements, their estimated density, and the number of sur-viving artefacts and burial grounds have all served as proxies. Adjustments are made for agrarian versus foraging societies, and scaling to larger areas is attempted to account for known human migrations. It gets better as we approach 'history', and our numbers allegedly grow to about 200 million by 1 AD. There is wide convergence of opinions that the population was largely stagnant for 1,000 years after, before it started to trend upwards. There are many accounts from then on, albeit patchy. Somehow, 1804 represents the year we reached one billion. Just over a century after, we added another billion. Becoming adept, we added four more before the 20th century was out. What's certain is that, historically, the explosion of the population is a recent phenomenon.

## Fast-Forward to a Mirage

Compelled as we feel to uncover our past, we're even more driven to peek into our future. At least our projected estimates are supported by a strong-enough base. Generally, projections can directly extrapolate high-level aggregates or detailed components, which will be then used as building blocks in subsequent aggregation. In either case, projections can rely on different growth scenarios resulting in bands. The data in Figure 3.1 from the US Census Bureau are based on projected populations for every country at a detailed level. Data from censuses, administrative, and survey sources are combined with relevant reports on public health, sociopolitical circumstances, catastrophic and other events to arrive at the best possible estimates for future fertility, mortality, and migration. Projected levels, trends, and patterns are then determined by the cohort-component method, using age and sex.[6] That's a lot of work. Is it worth it?

Population changes, both growth and decline, can have serious implications for societies. A heads-up is quite useful, particularly if it involves a bit more than a general sense of direction. Again, precision is not the imperative. Known unknowns will render assumptions wrong and surprises are bound to happen. Knowing that we're almost eight billion now, it'd be good to know if over the next 20–30 years we may stay the same, grow by a billion, or we peaked and are heading for shrinkage. The consequences of each scenario are different but for all they're huge. If national population estimates are indispensable for the functioning of a modern country (and drove policies such as China's one-child policy), this outlook ahead is critical for the world. Outside Malthusian concerns, and taking stock of our footprint on the planet, many key questions must be addressed. In the process, they could guide a multitude of decisions. With a billion more people, can we continue to rely on increased yields, if most of the extra food needed cannot come from new acreage? We've done quite a bit of that already. Is it wise to invest more in such research? What about housing and living norms? How much can already congested cities, airports, or train terminals expand before new ones are needed? Are more and bigger portables the way to expand classrooms? If we're headed for a billion less, what'd be best for the extra housing stock? Will rents collapse? What about downtown real estate and shopping malls? How many foreign language teachers can be sustained? An endless array of orthodox questions, heterodox opinions, and out-of-the-box fantasies will have data arenas to fight it out. Informed ignorance will beat uninformed arrogance.

That we're placing more demands on the planet is not news. Globally, higher fertility than the 2.1 replenishment rate makes our numbers grow, albeit currently the rate of population growth is decelerating. If ten billion of us can still be accommodated but 11 can't is ours to deal with. Whether our science provides homemade solutions, we mine resources from meteorites or something else, it won't be the first time we have to think about all that. A few decades ago, a running joke in statistical circles was that death may not be inevitable after all. (*"There are more people alive today than all the people who ever lived. Therefore, death is not a statistically significant result"*.) As if 'statistically significant' results need a sample bigger than what's left behind in the population from which it was drawn. But a serious answer is not befitting a joke. The beginning of this unfortunate shenanigan was probably a stir created when the UN started to produce population projections based on the new post-war censuses, including the 1953 one in China that pegged its population at 582.6 million. It was given a boost when global population reached three billion in 1960, earlier than forecasted. Classic Malthusian fears (we're becoming dangerously and unsustainably many), combined with concerns of the impact of human activity on the environment, raised red flags. Several studies debunked the myth that we outnumbered all our ancestors combined. The number of all people ever lived is now placed at over 117 billion.[7] Answering such questions also need estimates of longevity, which brings us to the next topic.

## Live Long and Prosper

The alien in our story asked how long earthlings live. Was the answer he got a good one? Life expectancy is one of those statistics that has risen to prominence and has become popular. As the data are on our side, it conveys a feel-good story about ourselves and our progress. Globally, from a lowly 47 years in 1950 it rose to 72.3 by 2020, with women now having almost five years on men. In many countries, life expectancy today exceeds 80 years. There are still big differences among countries but they have narrowed over time, as indicated by the minimum and maximum values of the two periods in Table 3.1. It's quite common for gains in life expectancy to be interpreted as humans now live longer lives. If I'm the one to break it to you, sorry…there's truth to that but it ain't quite so. Explaining in non-technical terms why life expectancy should not be confused with human longevity involves some nuances that need unpacking. We're here to get to know our numbers, right?

Life expectancies can be calculated for any group of people from *life tables* constructed with detailed data on deaths and survivors for each age (or abridged, typically in five-year bands). Life expectancy is a hypothetical average of all people in a population. Effectively, it's the sum of the probabilities of surviving from one age to the next, keeping mortality rates by age constant.[§] If death rates increase, actual life spans will be lower than life expectancy and vice versa. This is precisely what happened in the US and elsewhere after a year of COVID-19, lowering life expectancy. There are two forces at play simultaneously. The average length of people's lives at the time of death, as in the average age of people who die every year, is one matter.

---

§ An imperfect analogy would be the 3.5 expected value of the roll of a die (1/6 + 2/6 + 3/6 + 4/6 + 5/6 + 6/6). Computed figures for life expectancy vary somewhat across organizations because of different methodological choices and techniques.

**TABLE 3.1**

Life expectancy over time, by gender and age

| Life Expectancy | 1950–1955 | | | 2015–2020 | | |
|---|---|---|---|---|---|---|
| | Avg | Min | Max | Avg | Min | Max |
| At birth – All | 47.0 | 27.0 | 72.8 | 72.3 | 52.7 | 84.6 |
| Male | 45.5 | 26.3 | 71.0 | 69.9 | 50.5 | 81.8 |
| Female | 48.5 | 27.7 | 74.6 | 74.7 | 54.9 | 87.5 |
| At age 15 | 62.0 | | | 75.8 | | |
| At age 60 | 74.1 | | | 80.7 | | |
| At age 80 | 85.2 | | | 88.2 | | |

The question of longevity is another – think of the physiological limits of a human body.

Length of life is an individual experience. People die at any age, some at birth or very young, others at middle age, and some live very long lives. This has always been so, and still is. But, except for our higher absolute numbers today, the *proportion* of people who survive at any age is also higher than before. This is driven by decreases in mortality rates and is the main reason for the increase in life expectancy. Figure 3.2 shows that the probability of dying has declined over time but particularly at older ages and very young ages, thanks to progress in child mortality. This also explains why life expectancy at higher ages, when added to age, results in even longer lives (bottom part of Table 3.1). The 72.3 years of life expectancy for someone born today turn to 80.7 for someone already at the age of 60, and 88.2 for someone who's already made it to 80. Notice also that life expectancies at birth and at age 15 in the 1950–1955 period were identical. A newborn then was expected to live 47 years. A 15-year-old was also expected to live an additional 47 years, for a total of 62. In other words, the difficult part then was to survive up to the first 15 years.

A mind game in which we put a hypothetical upper limit to human life at some old age, say 90 or 100, would rule out the possibility of an increase in life expectancy from someone making it past that hard-stop. But life expectancy would still increase if the number of people who make it to the top increased or if they died at older ages than they did. Thankfully, there's no such hard-stop and we continue to push the limit.** So, in addition to how many people make it to old age, old age itself is shifting up a bit. This is longevity or life span, and it too affects life expectancy – but not as much.

---

** The record for the oldest person ever lived is held by Jeanne Calment from France at 122 years, 164 days (1875–1997). The oldest known person alive at the time of writing was Kane Tanaka from Japan, born on January 2, 1903. However, she died on April 19, 2022 at the age of 119. This made Lucile Randon from France, born on February 11, 1904, the oldest supercentenarian currently alive.

**FIGURE 3.2**
Probability of dying

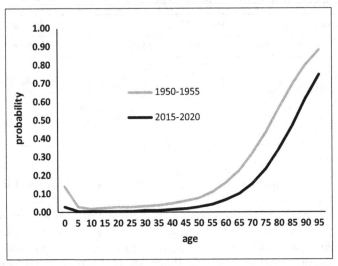

There's no doubt that there are more centenarians, golden anniversaries, and elderly with great-grandchildren alive today than ever before. But from Methuselah down, we have accounts of many people from the distant past living long lives. Surely, improved hygienic conditions, lifestyles, less hardship, and modern medicine have all contributed to buy us extra time at old age. Mostly, however, life expectancy has increased because proportionally more of us are alive at any age. To sum it all up, today:

- there are more people alive at every age, in absolute numbers
- there are proportionally more people alive at each age than earlier
- longevity has increased

The first fact is reflected in demographic statistics, the last two are captured by life expectancy.

## What's in Store?

Nowadays, knowing our numbers is crucial. As we saw, we do put in a good effort. But there are always sceptics. As we go through our daily lives the relevance of the population count is not as immediate as, say, that of the weather that may call for an umbrella. Sooner or later, you'll come across something along the lines: *"People didn't know such numbers for millennia and they were fine. Why do we need to know now"*? Admittedly, I've had such thoughts many times myself. Other than the need to define 'fine', trying to think of any answer makes you realize that this is a bad question! The same could be asked for the GDP, the CPI, and our monthly bills. For millennia we didn't have cars, banking, or cable television. No schools, hospitals, aeroplanes, or smartphones – one could hardly find a decent latte. Presumably, someone can always take it to the mountains but functioning in today's societies, let alone doing 'fine', has different requirements. Much better questions, and which I've heard repeatedly, are of the type: *"If such numbers are so important, why only count every 10 years"* and *"Is there no other way than a census"*?

It's already been mentioned that a census is a big, time-consuming, and relatively expensive exercise. Moreover, it's a means, not an end. The recommmendation for taking a census at least once every ten years was actually an aspirational target set by the UN in the post-World War II era. It reflected the state of our collective numeracy then. Still, many countries are not able to clear the bar. But much has changed in the intervening decades. The census, while authoritative, is no longer the sole source of population counts and related statistics. *Registers* for vital statistics and other sources used in intercensal programs continue to mature and can substitute for some census content. A few countries, notably Scandinavian, that have historically invested in good quality and interlinked *registers* of population, housing, employment, education, and more are already at the point not to need a census. Since 2004, France produces annual estimates by conducting a *rolling census* through surveys of geographical segments that cover the whole country over a five-year cycle. New thinking propels innovative efforts underway.

There is no underlying reason why the census cannot turn to a fondly remembered but archaic institution. Statistically advanced societies will rely on registers to know their populations from 'cradle to grave'. We shouldn't have to ask the same person when she was born, even every ten years. Our life events, important moments, and whereabouts are all recorded. Increasingly, our transactions leave behind digital traces too. Design improvements in information capture and coordination that facilitates its flows will enable the harvesting of the data we need. Moreover, this will happen in real time. There will be trade-offs, and choices will have to be made between the data that can be had from such an alternative system and any remaining needs that may require other means. In the process, countries that never quite took to censuses can leapfrog straight ahead, and direct investments accordingly.

However, the census will not go the way of the dodo anytime soon. It'll take much will, work, and patience to move to the future of data. In parallel with statistical matters, new thinking towards a legislative agenda will be necessary to revisit a host of issues, including confidentiality and privacy.

Inertia cannot be underestimated as a factor to slow evolution down. So, while you can already get a sneak preview of what will replace the census, don't book a museum visit to see an old census yet – but get ready to answer your next one.

## Fun and Games

Except for all the lofty policy and research uses, what else are all these numbers good for? What about some fun stuff? You bet!

Population numbers are used in all kinds of cool things. Online world maps and cartograms inscribe population on countries, in one colourful picture like a textured woven quilt. Artistic displays of country-balls depict the scale of their relative populations. Moving your cursor around them, more goodies pop up. There are also heat maps with rectangles for each country, in scale. (Not my favourite, the differences in scale are simply too big to work. With China and India taking up so much of the real estate, by the time you arrive at the bottom-right corner the rectangles are just dots.)

There are wonderful animations of human evolution over time that leave lasting impressions. In one, with a click on the start button, empty and dark-looking continents more than 2,000 years ago are transformed into very lit places, filled with bright-gold dots as the population explodes (much like some images of the earth from above, with night lights). Around 40,000 data points (200 countries and 200 years) and some code can get you in to see a visual production of a population race, with country-balls inflating against themselves and each other over the last two centuries. Neat! (Do check them out online.)[8]

The classics are still there, of course. Among the most successful early visualizations have been population pyramids – and for a good reason. They convey a lot of information in a simple, visually pleasing, and elegant fashion. Just one glance captures the distribution of a country's population by age, allows comparisons between the sexes at detailed age groups, and can even include absolute numbers for all. Some 'pyramids' are no longer pyramids. The name stuck because generally the age and sex structure of populations reflect the level of a country's socio-economic development. In developing countries, the base of the pyramid (younger ages) tends to be far wider than the top levels (older ages), reflecting both higher birth rates (which widen the base) and higher death rates (which narrow the top). As countries 'develop', they undergo demographic transitions under which both birth and death rates tend to fall, leading to age-sex structures bulging in the middle. The examples in Figure 3.3 illustrate. The one on the left is

**FIGURE 3.3**
Population pyramids

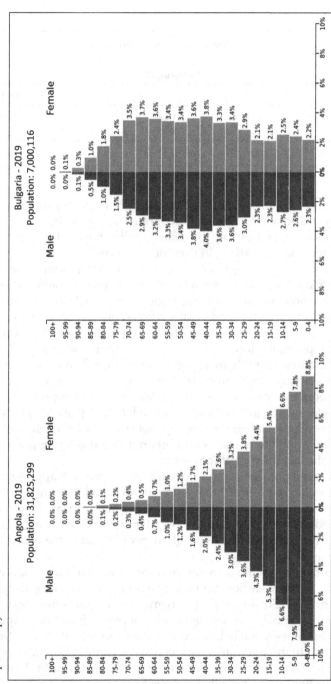

**Source: www.PopulationPyramid.net**

**TABLE 3.2**

Population metre

| World Population Metre<br>December 31, 2020, 11:59:59 pm<br>7,835,464,123 | | |
|---|---|---|
| **Today** | | **This year** |
| 386,537 | Births | 141,085,857 |
| 164,438 | Deaths | 60,019,950 |
| 222,098 | Net increase | 81,065,907 |

picture-perfect for climbing, the one on the right is wobbly and ascending to the top requires acrobatics. Population pyramids may not be in vogue in their static form nowadays, but they too come in fully animated versions showing rapid country transformations in front of your eyes.[9]

Just when you may have gotten the impression that it's a struggle to finish a census every ten years, or it takes a lot of work to produce annual population estimates, a whole bunch of real-time population clocks come to the rescue. Using crude rates for births and deaths, they come in many varieties, national and international, serious-looking or funny. Typically, they display more data than total population alone – such as the net additions in Table 3.2.[10] Since global population currently increases by more than 80 million a year (about the size of Germany), some quick arithmetic reveals that on average about 4.5 people are born every second, which explains why looking at the last digit of the digital clock can make you dizzy – even let some panic set in. (Much like the odd time you look at your electricity metre, realize how fast it rotates, and instinctively run in the house to turn off anything that's plugged.) I'm not a believer in the last three digits anyway.

With real-time population counters at hand, more cute apps have sprung up. You can plug in your birthday and get your pecking order in the global population, complete with trivia of happenings at that time. Someone born on February 29, 2000 would get 6,050,409,742, meaning that he can pull seniority to the almost two billion people that came after. Try it...if you're young.

Census archives contain a wealth of information on individuals and families that, as we saw earlier, are only released publicly many decades later to protect privacy. Such records, by themselves or in combination with any other source imaginable, enable genealogical research and have spurred the creation of businesses that help trace ancestors and recreate family trees. With a delay of two to three centuries, someone who came to Canada from Greece may be pleasantly surprised to hear from a long-lost 'cousin' living among vineyards in Bordeaux. (It really happened.)

## What Happened to 'The Good Old Days'?

You may have come across the statement *"Everything that can be invented has been invented"* – and more than once. Fairly or not, it's attributed to the commissioner of the US patent office in 1889. The supersonic jet, the smartphone, or the trip to the moon weren't even in the horizon then. Successive waves of technological breakthroughs brought marvels to our lives only to end up in museums. The 'TV' is not a furniture today but any screen you can connect online. The film, the VHS tape, the CD, and the clunky rotary-dial phone have all come and gone. We're now at the point of as-it-happens watching, at-will videoconferencing, and have instant fingertip access to masses of information and knowledge that, individually, we're not made with the capacity to process. Still, we know better than to follow the commissioner and declare the games over.

It's also true that most of our technological accomplishments happened in the 20th century, despite two world wars. The second half of the century was particularly transformative. We also know that much innovation happens incrementally, building on top of all past knowledge. In that sense, today's successes owe a lot to what came before. But still we don't seem able or willing to shy away from the due credit for what *we* accomplished in *our* times. We've outdone all those before. A whiff of superiority hovers somewhere in the air. A sense of *"we're so much better"* rubs off on us. Is this the answer to the question *"how did we manage all that"*? It's at this point that I'd like to call on the numbers we encountered in this chapter.

Populations experienced sustained increases throughout the period of technological expansion. Particularly in the second part of the 20th century, populations soared beyond expectations. During the same period, life expectancy increased by leaps and bounds. Could the temporal coincidence of the technological advancements and the fact that there are more of us and we live longer be a statistician's *spurious correlation*? The expanded workforce of today is not only due to our higher numbers. This is evident from its composition. Women entered massively the labour force during the same period, exerting an independent influence. Also, older people today are in better health and more educated than older people in the past. Not all go on golfing perpetually but many contribute to the workplace directly or through their associations.

Undoubtedly, strengthening the base of the pyramid diffused knowledge farther and wider than in the past. Practically, this has many implications. Increased specialization is one, going against the jack-of-all-trades moniker. Think of some popular movies where a (mostly) female city dweller returns to her small town where she meets a carpenter in the morning, who doubles as the hardware store owner in the afternoon, and reveals that he's also the mayor by the end of the day. We don't see much of that around us. We're also experiencing more distributed work processes, more team-oriented and less reliant on specific individuals. The old stereotype of a mad scientist toiling alone in a basement is laughed at. With all that in mind, I'd welcome more research on the effects population and longevity may have on the pace of technological progress and related socio-economic impacts. Who knows, it may even provide insights into our sustainable (or optimal) population size.

## A Proposal

This book is meant to explain numbers, it's not about proposals. But do indulge me this one time. If we're to make a serious move towards a register-based and real-time system capable of producing statistics from cradle to grave at will, a unique identifier of individuals is a must. It is the key to linking datasets at the microlevel, the level that matters. Such attempts struggle to a lesser or greater extent with multiple identifiers. China, many European, and other countries have citizen identification systems. India has also started one for its 1.4 billion residents.[11] North Americans rely on alternative, non-unique methods. Interactions with authorities and everyday business transactions are possible through wallets full of identification documents. Many are issued after a certain age, and you never quite know when you'll need your national ID, social insurance number, health card, passport, driver's licence, and a whole bunch more.

I propose that a number is assigned to each individual, at birth. The derogatory "*I was treated like a number*" doesn't cut it for me. I want to be a number. My own, unique, exclusive, and eternal number. *My* birth certificate, *my* first doctor visit, *my* school enrolment, *my* first or last job, *my* bank account, and *my* foreign border crossings will all be linked to me. Those with forms to be filled will not be asking me time and again for my name, address, date of birth, marital status, contact information and the like – only my number, *et voila*! They'll have it all.

Unfortunately, I can't tell you not to keep your maternal grandmother's maiden name handy. You'll always have to prove that you are you, and you're not a robot – even if the number I'm proposing was encrypted in a chip under your arm. Unique identification of individuals is a serious business. Just think of identity theft, biometrics, or AI for security – which works up to the point when a four-year-old is flagged at the airport for terrorism. Many people are working on such problems but, for statistical purposes, a unique identifier would go a long way. Some countries have already instituted similar systems for businesses, with unique identifiers for taxation and other purposes.

Depending on its actual implementation, such a number can be shorter than a long-distance telephone number. Twelve digits would be plenty, as it would take us to one short of a trillion. One idea is to make it the same as our order of arrival in this world. The number displayed by a real-time population metre at the exact moment of birth would do it, perhaps starting after 120,000,000,000 to account for all humans who've come and gone. If we want it to convey more substantial information upfront, it can be modelled after the date and time of our birth. So, 202011171015 would stand for November 17, 2020, at 10:15 am (computers will sort out the fact that, globally, there are currently 4.5 births per second on average). Smart people can easily design alternative, yet functional, national systems. For instance, three digits for the county of citizenship (like area codes), three more for the country of birth (if that's still of interest), two for the province or region of residence, even make it alphanumeric for gender or some other attribute. Then, 21395643027F could indicate a female Canadian citizen, born in Mexico, and residing in British Columbia.

If you think of something like that as far-fetched, consider the different numbers on the cards you now have in your wallet and how many times you're asked to provide your particulars. Then, think of the personal information you'd be willing to relinquish through your unique number. In addition to removing irritants and facilitating routine processes, this would be of tremendous help to the future of statistics – and, with no names.

---

## Fact-Checking Tips

*Know where to look:* A television commercial cunningly nudged people towards a financial advisor for the management of their investments by showing that they wouldn't dare to perform a surgical procedure on themselves. Some things are best left to the experts. From there on, though, it gets murky. Many times we value expertise and seek it out, other times we mock it. We easily concede that 'rocket science' is beyond our reach but we don't feel the same when it comes to our opinions on sports, socio-economic, or political events. Who's an expert, anyways, and what constitutes expertise? Other than putting in those 10,000 hrs in a specific area or activity,[12] what else is there? Leaving aside cases where special physical abilities or aptitudes are required, subject matter expertise in socio-economic research is rooted in deep immersion in a given area, which definitely extends to its numbers. Orders of magnitude won't do it, the bar is higher. An expert will know the numbers of her area inside-out. She'll also have intimate knowledge of all metadata, complete with details of original sources. Who are they, how do they produce the numbers, when do they revise them. A seasoned expert will also be able to see many real faces behind the numbers, she'll be networked.

No one can be an expert in everything. A polymath like Da Vinci in the 15th century cannot be in our midst today. Most of our fact-checking will inevitably revolve around numbers on which we're not experts. While the widespread advice to use orders of magnitude is still valid, there may be instances when it'd be of no help. As we discussed, there may be multiple and slightly different variants in circulation for what's supposed to be one number, for example, the population of a country in a given year. Only an expert will know (or may instinctively sense) and be able to explain away minor discrepancies among a yearly average, an end-of-year estimate, and another one for July 1 to satisfy international reporting requirements. Only she would know if the estimate used was revised the day before the actual publication of an already-written story. The author himself may not have been an expert either. In such cases, the numbers will be very close and any errors will likely be unintentional and benign. You need to rely on the context of the story and any background information available. Unless you determine a clear sinister intent, attacking with venom is not warranted. Leniency ought to be the outcome of fact-checking inconsequential errors. The use of numbers needs to be encouraged after all. Key questions to ask are: Does the story mean to inform through indicative numbers or provoke controversy through *any* numbers? Are the particular numbers, which may be a little off, central to the main findings or part of the supporting cast?

There's more you can do if the story is sloppy and the numbers noticeably outdated. Making a few remarks or raising good-faith questions in a private exchange with the author could prove fruitful. It may even lead to him doing the extra work needed. If you judge that it'd be too time-consuming to sort matters out due to lack of expertise, identifying and checking with the data source is a legit course of action. Intuition alone can take someone part of the way. If telecommunications data are at issue, contacting the Department of Justice shouldn't cross the mind. If data on farm irrigation and soil nutrients are concerned, the Department of Agriculture could be a plausible door to knock at. When in doubt, an inquiry to statistical outfits or an expert would be a reasonable first step. Eventually, fact-checking breeds long-lasting familiarity with data sources. Meanwhile, no need to take matters exclusively in your hands as the sole arbiter of truth.

*Predictions:* 'Fact'-checking predictions is an oxymoron. There are no established future facts against which to judge. Still, this doesn't mean that it's a free-for-all and that baseless exaggerations should be allowed to mislead or stoke fears with impunity. Estimates for some future point are produced through a variety of techniques. Straight-line predictions are quite common, others are based on regressions (most times also linear), or rely on the use of factors that scale some total. They may be presented as point estimates or as ranges, as when they're linked to alternative growth scenarios. Straight-line projections are simple and easy to follow, but they're less credible the farther away the time horizon gets. If pushed enough, they'll produce unrealistic or non-sensical results. Such would be the case of running 100-m dash races in negative time if the pace of breaking the world record continues. When GDP growth hovers around 2%–3% for decades, using a 7% growth rate without a convincing underlying reason won't work well. The too-good-to-be-true criterion applies. More than the projected numbers, fact-checking commonly looks at methodologies and the plausibility of the assumptions used. At a minimum, the audience should be adequately informed about *how* the predictions are made. Tossing up a number for ten years out with no explanation can't stand.

Data producers are weary of errors as they're direct hits on their most valuable asset, credibility. They tend to implement conservative quality assurance checklists to avoid mishaps. Some may be more daring, especially when any reputation loss won't be felt immediately but many years down the road. Nowhere are such 'poetic licenses' more pronounced than in market research, where unsupported claims may be based on methods not fully disclosed or have no reference to compare with. The minimum fact-check here is the classic orders of magnitude to separate the conceivable from pure fantasies. A few examples can come from serious-looking articles that exalt the benefits of upcoming technologies for global markets and point to the investment potential in companies behind them. Let's look at a few hypothetical ones.

  i.  *The smartphone that will replace all smartphones is expected to achieve 115% worldwide penetration by 2030.*

Smartphones are for individual use. Someone unfamiliar with telecom statistics will be well excused to think that this is bologna. Penetration rates among people can't possibly exceed 100%. However, such rates are historically computed with *subscriptions* rather than *subscribers* in the numerator. For a number of reasons, many individuals have multiple SIM cards – as the metadata explain. The International Telecommunications Union publishes such indicators for cell phones with values that exceeded 100% in some countries as far back as 2002.[13] While you may have doubts about this alleged success, the number isn't crazy.

  ii.  *This revolutionary VR social media platform acquired Facebook and plans an expansion that will support 10 billion accounts by 2025.*

Can that be? It's more than the forecasted number of people on the planet at that time. Now, Facebook claims a huge three billion users today. It's tenable that with amazing new functionalities the platform can hold on to all of them and still add another billion or so. With more optimism, and attrition among older people, it could conceivably go even higher. Then, the article refers to *accounts.* If the new platform allows accounts for made-up VR personas, this can be a game changer. Adding accounts for businesses, teams, and groups of people, who knows? Moreover, the statement refers to 'support' all those accounts. This is more of a supply-side and capacity measure than actual uptake. You may find the number highly unlikely but not in the outlandish sphere. You definitely need some clarifications and more information.

iii. *With built-in innovative features this all-in-one pitch-side Video Assistant Referee equipment (VAR) for soccer will dominate the market, which could reach 5 million units a year by 2030.*

You don't know much about soccer. What mental math can you do here? Starting with the device, you find out that after years of testing a video replay review was introduced recently, and FIFA took control of its implementation in 2020.[14] The 'all-in-one' unit is a monitor encased in protective gear and placed on a stand by the side of the pitch for the referee to check when needed.[15] You associate the unit with stadiums rather than matches. Now you have a target to quantify. A question to FIFA could help. Meantime, a very quick search will tell you that there are 37 big soccer *stadiums* in Brazil, which feels low. A more thorough search will reveal 187 soccer *venues* in that country,[16] which you round up to 200. You may allow even for backup units and generously imagine many more soccer fields in villages and neighbourhood parks where pee-wee games are played – although the use of VARs would be unlikely. Even if you push the number to a thousand or more, how many countries with Brazil's population and passion for soccer are there? Whether you multiply the number by 100 or 200 you're nowhere in the ballpark of the market size mentioned.

iv. *State-of-the-art vending machines onboard aircraft will deliver hot and cold food and beverages. Airlines can cut air crew in half. By 2025, the market will be worth $5 billion.*

Here you must contend with the future *value* of the market, for which you need to uncover prices, make assumptions about growth on air travel, the useful life of the machines, aircraft attrition, and other 'details' not provided. But it still helps to start with units. A quick search at the International Civil Aviation Organization (ICAO) will show that in 2019 there were 31,023 aircraft in service up from 23,880 in 2010,[17] and that the major manufacturers deliver about 1,600 aircraft annually in recent years. Whether from the stock of the aircraft or the annual flows, and assuming an overly optimistic growth while ignoring fleet attrition, you may arrive at a very generous 45,000 aircraft by 2025. Where would you peg the size of the 'market' in 2025? The article doesn't mention if the entire fleet is expected to be equipped with such machines by then, the shelf-life of the machines before retrofits are necessary, or much else. In the absence of necessary information, you may divide any 'reasonable' guestimate of annual sales into the $5 billion market. Chances are you'll end up with too stiff a price – even with machines that allow wireless ordering from your seat and other state-of-the-art touches! For no reason provided in the article, you can still cut any such estimate in half by allowing for two vending machines on each aircraft. Would that start to feel somewhat more fathomable? Could that be the 'market' the author was referring to? You shouldn't have to guess so many things and branch out deeper into a rabbit hole with less and less confidence. This onus here is not on fact-checking. The information provided has no legs to stand on. You can still buy stock in the company but not based on the information provided.

Where things take a really dark turn is when an outlandish projection is made for something that no one has even heard of. This happens around buzzwords in the marketing world. While it's possible that a new phenomenon will eventually catch, much early hype fades away. You can always take a shot at the projected size of the beef market in five years but when you see for the first time that the market for DfCaaS will increase from $2 billion today to $75 billion in 20 years, you can't. The acronym may stand for *Data from the Cloud as a Service*, but nobody has heard of that market – and no one has ever measured it. The same source crafts a number (itself suspicious) and then projects it! Such things need to be called out for what they are. The starting point is not the projection but the usefulness of the new construct and the validity of the present number.

---

## Notes

1   In Canada, a 'modified' *de jure* census is used. *"The population enumerated consists of usual residents of Canada who are Canadian citizens (by birth or by naturalization), landed immigrants and non-permanent residents and their families living with them in Canada. Non-permanent residents are persons who hold a work or student permit, or who claim refugee status. The census also counts Canadian citizens and landed immigrants who are temporarily outside the country on Census Day. This includes federal and provincial government employees working outside Canada, Canadian embassy staff posted to other countries, members of the Canadian Forces stationed abroad, all Canadian crew members of merchant vessels and their families. Foreign residents such as representatives of a foreign government assigned to an embassy, high commission or other diplomatic mission in Canada, and residents of another country who are visiting Canada temporarily are not covered by the census"*. The word 'temporarily' is intended to clarify the question of residence – do you live here or not? See, Statistics Canada, "Census of Population, Detailed Information for 2021," *Statistics Canada*, www23.statcan.gc.ca/imdb/p2SV.pl?Function=getSurvey&SDDS=3901

2 Statistics Canada, "Annual Demographic Estimates: Canada, Provinces and Territories, 2021," *Statistics Canada*, www150.statcan.gc.ca/n1/pub/91-215-x/2021001/tbl/tbl2.1-eng.htm

3 US Census Bureau, "International Database: Population Estimates and Projections Methodology," *US Census Bureau*, www2.census.gov/programs-surveys/international-programs/technical-documentation/methodology/idb-methodology.pdf

4 "…*instead of using only comparative and superlative words…express myself in terms of number, weight or measure…*" Excerpt from quote in I. Bernard Cohen, *The Triumph of Numbers: How Counting Shaped Modern Life*. Norton, 2016.

5 The estimates are either as they appear or slightly modified averages from several sources, including the US Bureau of the Census – which, in turn, credits several sources, including Ralph Thomlinson (*Demographic Problems, Controversy Over Population Control*, 1975) and Colin McEvede and Richard Jones (*Atlas of World Population History*, 1978). Others cite John Carl Nelson (*Atlas of the Eight Billion: World Population History 3000BCE to 2020*, 2014) and Massimo Livi Bacci (*A Concise History of World Population*, 2017). More recently, the so-called Cologne protocol is used as a 'mathematical interface between epochs', to fill gaps in whole mitochondrial genomes' DNA and genome studies that examine pre-historic populations.

6 US Census Bureau, "International Database: Population Estimates and Projections Methodology."

7 This is based on a much-referenced study carried out by Carl Haub from the Population Reference Bureau in the US in 1995. Our population then accounted for about 6.5% of that, but as such studies continue to be updated it inched up to 7%, and 117 billion, by 2020. See Toshiko Kaneda and Carl Haub, "How Many People Have Ever Lived on Earth?" *Population Reference Bureau (PRB)*, www.prb.org/articles/how-many-people-have-ever-lived-on-earth/, May 18, 2021.

8 Some can be found at www.gapminder.org/fw/world-health-chart/ created by the late Hans Rosling. Also at https://worldpopulationhistory.org/map/2050/mercator/1/0/25

9 See for instance, Population Pyramid.net, "Population Pyramids of the World from 1950 to 2100," www.populationpyramid.net/

10 Many examples exist. For instance, https://weta.org/watch/shows/humanity-space/ humanity-space-population-clock, www.worldometers.info/world-population/, https://worldstatistics.live/population, www.livepopulation.com/

11 In recent years, India intensified efforts to roll out a voluntary identification system (known as Aadhaar). Initially launched in 2009, a 12-digit unique ID aims to eliminate multiple and fake IDs and is linked to the residents' basic demographic and biometric information (photo, fingerprints, and iris scans). China has a mandatory ID for citizens over 16. With 18 digits, it captures place and date of birth. Such IDs are required for many activities. Unique Identification Authority of India, Government of India, "myAadhaar, one portal for all online services," *Unique Identification Authority of India*, https://uidai.gov.in

12 Malcolm Gladwell in *Outliers: The Story of Success* (Penguin Books, 2011) claimed that this is the number that's key to success in something. Intensive practice to achieve mastery of complex skills and materials.

13  Data on mobile cellular subscriptions from the world communications/ICT data-base, www.itu.int/en/ITU-D/Statistics/

14  See Sport Performance Analysis, "Application of Video Technology in Football Refereeing – VAR," www.sportperformanceanalysis.com/article/application-of-video-technology-in-football-refereeing-var, November 6, 2019.

15  FIFA, "Video Assistant Referee (VAR) Technology," *FIFA*, www.fifa.com/techni cal/football-technology/standards/video-assistant-referee

16  A quick search for Brazil will show 37 stadiums with capacity more than 30,000. Wikipedia contributors, "List of football stadiums in Brazil," *Wikipedia, The Free Encyclopedia*, https://en.wikipedia.org/w/index.php?title=List_of_football_sta diums_in_Brazil&oldid=1080168076

17  ICAO, "Annual Report 2019," www.icao.int/annual-report-2018/Pages/the-world-of-air-transport-in-2018.aspx (citing Reed Business information). Commercial transport fleet of ICAO members includes turbojet and turboprop aircraft. "*Active and parked aircraft are included; aircraft having a maximum take-off mass of less than 9,000 kg (20,000 lbs) are not included*". Two-seater CESNAs won't need a vending machine.

# 4

## The Socio-Economic Realm

We discussed counting and measuring in the first two chapters, while Chapter 3 showed how counting turns into estimation. Well, this chapter is all about estimation.

The socio-economic realm is all about us and our lives. This realm is in constant interaction with the physical world through a series of 'transactional exchanges'. We extract, transform, and use resources, intervene and change the landscape, affect the environment. The socio-economic realm evolves over time and creates our *reality*. Ours today is different from that in feudal times. Billions of us are active, do things. The collection of our activities vis-à-vis the physical world and the interactions among ourselves create our *economy* and our *society*. Frequently, we connect them with a hyphen because it's *us*, the same people in both.

Our measurements don't have an independent existence. We invent them for our purposes and they evolve to serve the needs of our times, not unlike the Sumerians did long before. But our affairs have become more complex, our economies and societies more sophisticated, and our measurements increasingly abstract. Removed from the physical and the visible, notions and constructed ideas are the name of the game. Social sciences drive their quantification. Theory becomes prominent, relationships are explored through hypotheses, and explanations are sought through metrics. We continue our determined efforts to quantify what we experience, as well as where we want to steer that experience to. None of the statistics we have today existed in the past.

### The Rate of Unemployment

Years ago, I was off to a class for an economics course. That particular class was on the labour market and unemployment. I could have gone straight to the blackboard, but not that day. I opted for class engagement. Concepts stick better if students are part of the discovery process, I'd bring them along. Walking into the big auditorium with the hundreds of students, right after testing the mic by greeting the class, I began with *"Let's do some thinking*

DOI: 10.1201/9781003330806-4

*together*". Anticipating the answer to my first question, I phrased it in the negative: "*Who has **not** heard of the rate of unemployment?*"

I was right, not a single hand was raised. (I suppose a raised hand didn't represent an opportunity to shine in this case.) "*It figures*", I thought. This statistic is reported so widely and for so long that, regardless of how much someone knows *about* it, everyone knows *of* it. "*So, what do you think it is?*", I continued.

Many hands were raised. Those were the days...We heard a couple of circular opinions like "*Unemployed are those who can't find work*" until the next student started saying: "*It's the number of unemployed as a percentage of...*" before he paused. His hesitation lasted a second or two before he proceeded to add "*...the population*".

Well, not a bad start only a couple of minutes into a three-hour class, I thought. 'Percentage' responds well to 'rate', definitely in the right direction. I didn't have to say much for a few more minutes, the class took over. "*I think it's the percentage of the adult population*" said the next student. "*They don't count kids*". "*Adults are over 18...*" said another "*...but kids can have jobs before then*". Soon we sorted out that the legal working age in Canada is 15. But hands stayed up. "*I don't think it's all adults or those over 15*", said someone. "*What about seniors?*" That too came across as reasonable to the class, I could sense. Quick exchanges on who's *senior* weren't conclusive. Pensioners were mentioned, and those over 65 were mentioned too.

Still, so far, no home run. At least not among those I'd picked from those who'd raised their hands. But many hands were still up. One was waving at me particularly eagerly. I nodded a go-ahead. "*It's the percentage of the labour force*" she went on to say. "*Age doesn't matter, it's who's in the labour force that counts*". Either because they sensed a superior argument or because that's what they wanted to say too, many hands were lowered.

This pedagogical approach worked, at least in that class at that time. Crowdsourcing the answer implicated the students in the construction of a concept in a way that helps both understanding and retention. Moreover, the process front-loaded some questions that would have to be asked later anyway. In a matter of no more than a few minutes the students had retraced the thought process of the pioneers who constructed this statistic. They too went through this thinking – and much more.

I could have continued. I could have asked: "*What's the labour force?*" But that was the time I turned my back and wrote with white chalk on the huge blackboard (green, really):

$$u = U/LF = U/U + L$$

As far as mathematical formulas go, this is a lightweight. Nothing to it. The rate of unemployment (u) is equal to the number of unemployed individuals (U) divided by the labour force (LF), which in turn is the sum of unemployed and employed individuals (L). Typically, it's also multiplied by 100% to express it as a rate, but that's totally mechanical. Still, as I was writing down the last couple of terms, I could hear rumblings in the air behind me. I could sense unease, caused by the definition of the labour force and the inclusion of U in both the numerator and the denominator. The class was far from over.

## Give Me a Theory...

...and I can measure anything. There is such a saying. A theory, big or small, is behind most of our measurements in the socio-economic realm. The classic beginning is the invention of an abstract notion, in pure form, which intends to capture something we experience in real life. Later, it will be quantified.

There's no statistic that better straddles the economy and the society than the rate of unemployment. It's at the heart of the economy because of the labour market. It's also been called a social evil. Using this statistic as a guide to look at many issues involved in estimation in the socio-economic realm is quite apt.

There was no such thing as the rate of unemployment in earlier human history. In fact, the term *employment* came to use after the industrial revolution. Disregarding some ambivalent descriptions in earlier times, its first use wasn't before the end of the 19th century. Until then, people didn't identify with *unemployment*, and governments weren't into full-employment policies. Estimates started to appear in the 20th century in the US before the Great Depression, when unemployment became its unfortunate star. Post-World War II, many countries needed that kind of information to guide their transition from wartime to peace economies, and it was then that today's definitions were developed.

The concepts of employment and unemployment are grounded in the economic theory of the labour market. Like in every market, demand and supply determine equilibrium quantities and prices. In that setting, the price is the wage rate and the quantity is the number of employed individuals. Shifts in the demand or the supply of labour don't cause unemployment, equilibrium quantities and prices simply change to 'clear' the market. Unemployment emerges when prevailing wages are higher than the equilibrium. Then, there are people left involuntarily without work. (We also know that equilibrium doesn't mean *full*, *fair*, or *optimal* employment, right?)

Theory can go as far as to identify employment as a factor in the production of goods and services, leading to another notion – that of economic output. In that context, the notion of employment differs from *work*. For instance, unpaid housework or volunteer work are not labour market activities even though they don't differ from paid work. This is as far as theory would go. It doesn't prescribe how any of these concepts should be measured. It definitely doesn't venture into the role of surveys as a means to do so.

A particular strength of empirical statistics is to bring abstract concepts to life and, in the process of measurement, down to the real world. This applies to all new concepts, sometimes even flavour-of-the-month buzzwords. The thinking is not complete without drilling down, and there's always much to ask. What exactly does this mean? Who should be in, who should be left out? What is the intended use of this measurement? As we saw in the early chapters, there are reasons why humans want to measure something they know, or believe, is 'out there'. Invariably, the focused thinking forced by empirical measurements uncovers elements that were missed or overlooked, something that feeds back to theories and refines them. On the particular issue of the unemployment rate, statisticians had a lot more sway than usually as measurements happened in parallel, or even preceded, theoretical advances. On this very topic, David Card states that: "*Surprisingly, it was not until 1940 that our current conception of the labor force – and the equating of unemployment with active job search – finally emerged. The birth of the modern definition of unemployment represents a remarkable triumph of practical measurement needs over persistent concerns about the absence of theoretical underpinnings. Economists continue to remain skeptical of any single theoretical definition of 'unemployment'*".[1]

### Kicking the Can

A key attribute of estimation in the socio-economic realm is the difficulty to nail down conceptual constructs, which are well understood at some higher level. More often than not what happens is kicking the can down the road. Something is defined in terms of something else, which in turn needs defining, and so on. The students in the class walked down this lane when they started pondering 'details'. The definition of the *unemployment rate* defers to the definition of the *labour force*. This defers to the definitions of *employed* and *unemployed* individuals. Now, how hard could that be?

As in the case of the census, this exercise too involves counting people. Decisions concerning *whom, when, where*, and *how* to count are needed. Once again, they'll be guided by the intended use of the measurement. Ultimately, the definition arrived at will reflect the interplay between the theoretical scope of the concept and the use it'll be put to.

The idea to measure the unemployment rate was not driven by the desire to add one more conversation topic during happy hour. Presumably, the government needed to know so that it can monitor for worrisome signs and perhaps take actions for something so important to the well-being of the country. Also presumably, actions would be proportional to the magnitude of the problem, for which regular measurements would be necessary. Any policies for unemployment would obviously focus on the population of interest inside the country, which needs to be determined. Labour laws with minimum legal working age requirements impose a low threshold on the in-scope population.* The high-end threshold poses more of a challenge, but it's manageable under the logic of the intended use. Issues concerning retirees will be taken care of outside the labour market. Although countries have statutory retirement ages, many people receive pensions well before that. At the same time, older folks can be working at any age. So, even though not as clear-cut, we get an upper boundary too. The bulk of people roughly between the two boundaries (ages 15–65) is commonly referred to as *working age* population. Before we return to the *whom* should be included in the measurement we can expedite the other interrogatives.

Remember the census serving as a frame for other surveys? Well, in practice, statistics on the unemployment rate are produced by such surveys. The specifics differ from country to country but *labour force surveys* are typically good-size undertakings, with probabilistic samples stratified to produce data for many groups of people and geographic locations. They're also sub-annual, and in many countries monthly. They take place during a specific week each month – the *reference period*.† The resulting 'monthly' figures refer to that week, really.

There are additional exclusions from the scope of the survey for all kinds of reasons. They depend on the country but commonly exclude members of the armed forces, institutionalized populations (penal institutions, asylums, hospitals), and remote, hard-to-reach populations.‡

Then, employed are defined as those who during the reference week:

- had paid employment or were self-employed
- did unpaid work for a family business

The definition of the labour force is now down to the definition of the unemployed. This proves trickier, certainly less direct. Surely, someone must be out of work during the reference period. But many people are. Whether stay-at-home parents, independently wealthy individuals or others, many non-working people are not seeking employment and therefore are not

---

* In the case of a survey, the population of interest is determined through a combination of its scope (i.e., excluding under-age children) and the information collected (i.e., retirees). In Canada, the low threshold is the age of 15. The US collects data for age 15 and older but only publishes them for over 16.

† To get a peek into how the statistical system works, labour force surveys also serve as frames for smaller surveys. These piggy-backs take advantage of the monthly infrastructure, inserting questions of interest for fast turnaround either in the whole sample or parts of it. In Canada, the labour force survey is monthly since the early 1950s. It determines the status of respondents based on their activities during the reference week, which *"usually contains the 15th day of the month and stretches from Sunday to Saturday"*. In the US, it's the 12th of the month. In both Canada and the US, pre-announced releases of results are scheduled in the first Friday of the month after the reference month.

‡ In Canada, these groups represent an exclusion of approximately 2% of the population aged 15 and over.

part of the intended use of the measurement. To weed them out, the defin-ition requires that individuals must be *available* and *actively looking* for work. Clearly, these criteria are open to some interpretation. Now, they need to be defined. Another kick at the can.

*Available* disqualifies full-time students during the academic year and 'returning students' between academic years. It also provides further justifi-cation to earlier exclusions from the labour force. Then, waiting at home for someone to knock at the door doesn't cut it for *actively looking*. Depending on the time and place, someone may physically go out for job-hunting, check classified ads, search online, make calls, or ask family and friends. The survey asks probing questions to determine the labour force status of indi-viduals. Practically, the line is mostly drawn with respect to the past four weeks. Unemployed is someone who didn't work during the reference week, was available for work, and either:

- had looked for work in the past four weeks or
- was on temporary layoff with a reasonable expectation of recall or
- had lined up a job that would start within four weeks from the refer-ence week

(The last two categories don't actually require active looking.)

So, the seemingly simple task of allocating the in-scope working-age popu-lation into mutually exclusive groups of employed, unemployed, and not-in-the-labour-force is accomplished through a lengthy series of steps involving more and more detailed information. Decisions made along the way represent a balancing act between the theoretically sound and the practically feasible. There are many people and many stories. The consistent application of cri-teria can lead to conflicts that force choices on which side to err. These create ambiguities or even 'collateral damage' with respect to the intended use of the measured statistic.

A key example comes from a group among the unemployed who not only hadn't looked for work in the past four weeks but, even worse, had given up looking. Consequently, they're not counted as unemployed. There is in fact a specific term for them, the self-explanatory *discouraged workers*. You'll justifiably come across this as an argument that the official rate of unemploy-ment underestimates the true rate. No doubt, it does. The problem is bigger in bad economic times, when the prospect of finding work is at its lowest and the number of discouraged workers at its highest. But thanks to probing questions their number is estimated, so there's a way out.

Then, there are those who see value in a more elastic application of the *actively looking* criterion. For instance, the Bureau of Labour Statistics (BLS) in the US makes discouraged workers a sub-set of a broader group of indi-viduals. *"Persons marginally attached to the labor force are those who currently are neither working nor looking for work but indicate that they want and are available for a job and have looked for work sometime in the past 12 months".*[2] This leads to an even higher unemployment rate. Time also enters the thinking, whether it refers to the duration of unemployment or the hours worked. Short spells of unemployment, such as when moving between jobs, are qualitatively different from chronic unemployment. The duration of unemployment, under or over 15 weeks, is also measured in the US. Then, there are those who'd like to work more hours than they do. Part-timers who couldn't find full-time work and full-timers who couldn't work their full hours. Collecting such data makes it possible to estimate *underemployment* too.

All these 'corrective actions' lead to a cornucopia of different measurements. Inquisitive minds followed a path that has led to an embarrassment of riches. Rather than succeeding in a clear definition, we end up with a menu of unemployment rates next to the official one.[§] The difference from the narrower to the broadest can easily be a factor of three or four.

Documenting all that brings to the forefront again the importance of metadata. I hope that after this explanation they're no longer perceived as a technical and nerdy term but an absolutely essential companion of any measurement. Metadata explain, clarify, and qualify what a statistic actually measures and what it means, not what someone thinks it measures, should measure, or ought to mean.

## A Stylized Approach

By the time we arrive at a full comprehension of the numbers needed to feed the simple formula of the unemployment rate, several important aspects of the labour market have surfaced. The 'truth' is not only what comes out of that constructed concept but also what's ignored or concealed by it. Clearly, the unemployment rate is but one member of a larger family of complementary metrics. An illustration will solidify their identities and their different personalities. As relative magnitudes differ across countries and over time, we'll rely on a stylized example from a hypothetical, yet plausible, economy. Figure 4.1 and Table 4.1 serve as props.

Starting with the entire population of a country (the whole rectangle), those under the legal working age are sliced out-of-scope (*a*). So are retirees on the other end (*g*), in a way unrelated to their age. Retirees can be younger than age 65, while working individuals can be of any age. The labour force

---

§ The BLS calculates the following statistics: U1, the percentage of labour force unemployed for 15 weeks or longer; U2, the percentage of labour force who lost jobs or completed temporary work; U3, the official unemployment rate that occurs when people are without jobs and they have actively looked for work within the past four weeks; U4, the individuals described in U3 plus discouraged workers; U5, the individuals described in U4 plus other 'marginally attached workers', 'loosely attached workers', or those who would like to work, are able to work, but have not looked for work recently, and; U6, the individuals described in U5 plus part-time workers who want to work full-time but cannot due to economic reasons, primarily underemployment.

**FIGURE 4.1**

A stylized view of the population and the labour force

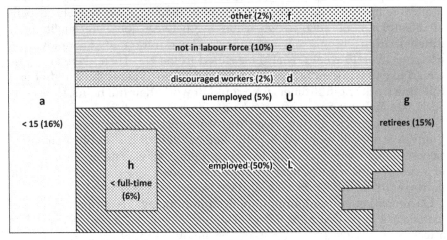

comprises the employed (*L*) and the unemployed (*U*). Discouraged workers (*d*), others excluded from the labour force (*e*), and employed individuals unsatisfied with their working hours (*h*) are also shown, as are populations excluded from the survey scope altogether (*f*).[*]

Table 4.1 shows how data from a labour force survey can be combined with additional demographic data, in absolute and percentage terms, to construct many indicators of interest. A series of unemployment rates is possible, but also indicators such as the *employment rate*,[††] the *participation rate* (labour force as a proportion of working-age population), and the *dependency ratio* (non-working age population as a proportion of working-age population). By now, you know that you can create your very own too! A little favourite of mine is quite simply the proportion of the employed in the total population – in our stylized example a full half! It crossed my mind many a time on the way to work.

When the students weren't at ease with the inclusion of U in both the numerator and the denominator, they were really toying with different possible expressions that can capture an unemployment rate concept. We now know

---

[*]  The dotted lines between the 'other' category and retirees or those under 15 indicate 'porous' borders, with possible overlaps between groups. So, *other* refers to additional individuals.

[††]  Employment rates can be expressed as employment against the labour force (exactly like the unemployment rate), the proportion of employed individuals over the working age population, or working individuals of a certain age as a proportion of the total adult population. In principle, both the numerator and the denominator can be adjusted, as it was done in our example for working individuals past the age of 65. Retirees under 65 could have been added to the denominator too.

**TABLE 4.1**

The rate of unemployment and related statistics

| Population | Millions 50.0 | (%) 100.0 | Statistic | Definition | % |
|---|---|---|---|---|---|
| <15 (a) | 8.0 | 16.0 | u rate | $u=(U/L+U)\times100$ | 9.1 |
| 15–65 (b) | 33.0 | 66.0 | | | |
| >65 (c) | 9.0 | 18.0 | u(d) rate | $u(d)=(U+d)/(L+U+d)\times100$ | 12.3 |
| >65, still working (k) | 1.5 | 3.0 | | | |
| Labour Force (LF) | 27.5 | 55.0 | Underemployment rate | $u(r)=(U+d+h)/(L+U+d+h)\times100$ | 20.6 |
| Unemployed (U) | 2.5 | 5.0 | | | |
| Employed (L) | 25.0 | 50.0 | Employment rate | $l=(L/b+c–k)\times100$ | 61.7 |
| full-time | 20.0 | 40.0 | | | |
| <full-time (h) | 3.0 | 6.0 | Participation ratio % | $p=(L+U)/(1–a)\times100$ | 65.5 |
| self-employed | 5.0 | 10.0 | | | |
| Retired (g) | 7.5 | 15.0 | Dependency ratio | $d=(a+c/b)\times100$ | 51.5 |
| Not in LF (e) | 5.0 | 10.0 | | | |
| Discouraged (d) | 1.0 | 2.0 | Employment/population ratio | L ratio = $(L/population) \times 100$ | 50.0 |
| Other (f) | 1.0 | 2.0 | | | |

that such expressions actually exist and are useful. What's more, research not only permits but demands the exploration of alternative paths. What if instead we simply used U/L? The resulting number would be different from what we're used to, but it would move just the same and will do just as good a job.

One more thing. The mathematics of the unemployment rate formula makes it possible for the number of unemployed people and the unemployment rate to move in opposite directions. This is neither magic nor chicanery but a direct corollary of the definition used. How many times have you heard arguments from opposite sides, 'supported by the data', on how well or how badly the labour market fares through convenient picking of absolute or relative numbers particularly if comparisons are made over long periods? But people can get away with this only if the audience is uninformed. Numeracy again can act as a reality check.

## The Makings of an Expert

Despite all the hidden detail that the seemingly simple rate of unemployment contains, it's not as remote or esoteric a statistic as, say, the money supply, the primary budget deficit, or another lesser-known notion in the economists' toolkit. Unlike the deference most people show to matters of theoretical physics, quantum mechanics, or rocket science, most anyone with a foggy idea of the unemployment rate feels free to step in and pontificate. You can call it *the people's statistic*.

To gain some distance from this, perhaps height, ever-resourceful economists have made sure that there's a lot more in their bag. Unemployment can assume a great many forms. It can be frictional, structural, cyclical, seasonal, 'natural', 'non-accelerating inflation', and more. Policy-relevant subcategories also emerge, claiming their right to be measured. These include broad aggregates, such as paid employment vs. self-employment, public sector vs. private, full-time vs. part-time, voluntary vs. involuntary, among others. They're also all ripe for dissection by gender, age, industry, lower level geography, occupation, and more. For each of them, we can compute rates of unemployment and other indicators among those we encountered.

To top it off, theory really asks that the logical unit for the measurement of the labour supply is not people but *person-hours*. This is intimately linked to productivity (but I'll spare my readers that pit). Truth is that people can be moonlighting, working at two or more jobs simultaneously. This is getting more pronounced in today's gig economy. Others work overtime. Moving from people to hours worked as the unit of measurement requires even more detailed data.

By the time someone masters the theory, its conceptual constructs, and their empirical implementation through survey work, the fun is about to continue. The end formulas are agnostic to the sources of their data. They don't care, they can ingest data regardless of where they come from. What's so sacred about a labour force survey? It's a self-reporting survey after all. Plug in data from the unemployment rolls, payroll information, business surveys, or any other source. With each choice, pros and cons will appear and trade-offs with surface. For instance, with regard to the unemployment rolls, not every unemployed may be eligible for benefits, some eligible may not be there, there may be time lags involved and other issues. One more practical example of difficulties to arrive at what seems like simple counts of people.

Further proof that some groups of people are difficult to measure with certainty comes directly from businesses. Every time they're asked for the number of their employees they cringe, they don't quite know. Between core employees, part-timers, seasonals, students, or contract workers that come and go, the number of employees in large businesses depends on the day! (Which is why surveys at times go after a specific day, e.g., the last day of September.) This has also given rise to the notion of FTEs, converting employees to *full-time equivalents* rather than actual bodies. On the other end of the spectrum, the matter is also hard to pin down. All economies have a very large number of small businesses, among which micro-enterprises. From incorporated legal entities to unincorporated self-employed individuals, they may hire occasional help, say, a summer student or two. Such practices do a number on business registers that don't quite know if those businesses have zero, one, or two employees – they're called *jumpers*, as they can change frequently affecting business demographic statistics and sampling for surveys with interest in such businesses.

So, a researcher must contend with a fair amount of theories, their empirical application, and a multitude of data sources with their intricacies. These are all linked but don't always talk directly to each other. Subject matter knowledge and patience are necessary to navigate through and understand issues ranging from the purity of concepts to the practical nitty-gritty of the scope, coverage, and frequency of their actual metrics to the technocratic implementation, monitoring, and evaluation of policies.

Finally, international comparisons are often needed in research. In principle, for the subject matter of labour markets and the unemployment rate, most countries adhere to standards by the International Labour Organization (ILO). Once again, however, "*...the guidelines are, by design, rather imprecise, so that individual countries can interpret them within their particular labour market context. As a result, unemployment rates are not strictly comparable across all countries*".[3] For instance, studies specific to the measurement of the unemployment rate between Canada and the US have concluded that the Canadian method of calculation adds a full percentage point to it.

Repeated exposure to all aspects of a subject matter area breeds fluency. Soon, you know what questions to ask. You become an expert! Now, beware of the paparazzi.

---

## Inside Empirical Research

Most socio-economic measurements are based on constructed, human-made concepts. As a result, they 'exist' only within the confines of their creation. Moreover, they're invented to serve specific needs, which may be repurposed, and defined with different degrees of precision. How all that happens is explained below.

### A Metaphor

Picture a middle-aged couple standing in the middle of a big and empty lot. There's nothing around them but land and air. They open a roll of blueprints and look intently, their eyes shifting back and forth between the plans and the ground, as if they measure. They point their fingers to a nearby spot and move a few steps. Then, they look at their feet and lift their heads slowly, gazing upwards. They quickly pass their eye level and keep going. And going. Their necks are now at full extension, looking at the sky.

This is not yoga or tai-chi. They're trying to mentally locate a spot 33 m high, at a vertical distance from their feet. Not easy without any points of reference around, like buildings or tall trees. According to the plans, this is where their 11th-floor condo will be located in a new high rise. Their bedroom will be about 5 m to the left of the front entrance they're trying to locate, which will be 6 m from the elevator. Their bed will be facing south, the plan shows, and the wide bedroom window will be to their left. They'll be able to see the sunrise from the east if they wanted. But they'll have to go to the balcony on the other end for sunsets. How can they possibly visualize the view out of the kitchen window? Perhaps, they'll bring a drone next time.

The couple mentally carved a piece of vacuum, thin air, to 'see' inside their bedroom and outside from their bedroom. They will soon spend a big part of their lives there, about a third. The architect and the engineer did the same before them. In fact, all our bedrooms fall exactly in the same category. Whatever floor yours is, the space used to be occupied by thin air before it was built.

There are analogies with the measurements of our human affairs. We borrow from the physical world and erect structures in a vast and empty conceptual space. We design and build all kinds of unique places, down to their detailed elements and finishing touches in the service of some purpose, some need. We plan meticulously, et voila! To serve our human activities, we invent notions and measurements that claim a piece of the conceptual real estate where previously nothing existed. Claiming, naming, and using such pieces of conceptual real estate is the name of the game in socio-economic research, and our way to understand our world and our lives.

Using the parallels of this analogy we can explore how this is actually done, and understand how abstract ideas morph into their applied form. We'll also codify the basic steps across all types of estimation and understand the actual production of existing measurements.

## Mental Creations

It all starts with mental creations that we call theories, or sometimes frameworks. A theory is an abstraction that establishes relationships between notions of interest, causal or not. Frequently, simplifying assumptions remove complexities that in the real world obscure the transmission of signals. A framework is a simpler way to establish relationships among notions, simply by postulating a 'logical' sequence connecting one to the other. Occasionally, notions can be just buzzwords describing some influential activity or phenomenon. Notions are sometimes called *variables*, albeit the same term is used in their applied form too. The term itself is a cry for measurement. We're always interested in measuring something that varies. However, the quantification of notions is only a back-of-mind concern at this stage. Theories, frameworks, or buzzwords and the notions they contain are shown as block A in Figure 4.2.

**FIGURE 4.2**

From theories to estimation and analysis, and back to theories

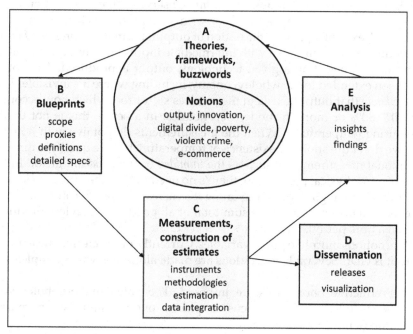

Theories and frameworks per se are not subject to measurement. However, the notions they contain are measurable and the hypothesized relationships among them become subject to testing. Theories live or die on the basis of their acceptability, which in turn depends on their explanatory power and ability to predict. Throughout, the *ceteris paribus* applies. For instance, an increase in the supply of money will eventually lead to lower interest rates and inflation – *other things equal.* For measurement purposes, the research focus is squarely on the notions, not the theories themselves.

Notions come in an abstract and pure form. They're not overly concerned with practical matters. Following our metaphor, theories and their notions are like drawings, an artist's rendition of a new structure. But nothing can be built from such a sketch. In the physical construction of the condo building, the 11th floor unit doesn't just appear. It sits on top of other floors and they're all supported by a strong foundation and advanced engineering. The development of detailed blueprints necessary for the actual construction is shown as block B. From decisions involving the scope of a notion to its definition, its eventual refinement and different variants, and the data needed for its measurement, much of the heavy lifting is done there.

## Output and GDP

Take, for instance, the notion of *output*. Observing that workers and raw material enter the factory gate and some finished product is loaded on trucks at the back exit fully justifies the notion of output. Formally expressed, factors of production are combined with intermediate inputs and, through a production function based on a given technology, output is produced. The notion was soon extended to the whole economy, including where it's invisible. That is, we *know* that output exists in the services sector too, which now accounts for 70%–80% or more in modern economies. In a sense, this is not unlike measuring temperatures. When the measurements do not jive with how we *feel*, we don't question the existence of temperature but we qualify it through additional measurements, such as the *humidex* or the *windchill* factor. In the case of non-physical production, there's no need to 'see' the output of service industries or the government to know it's there. The theory can go as far as to conceptualize output as the sum total of all goods and services produced within a time period.

Pure notions morph to proxy variables. Currently, our preferred measure for output is GDP. Several key decisions are made along the way. Examples are:

- Productive boundaries are imposed. The value of household work is excluded and the focus is placed overwhelmingly on market transactions.

- The notion of output refers to all goods and services in some quantitative sense but avoids the issue of aggregation, which is not trivial. The classic example from Economics 101 is guns and butter. Starting from an initial mix, what happens to overall output when one increases and the other decreases? Imagine aggregating across millions of goods and services today…The way out is to make output equal to the *value* of all goods and services produced. To stay faithful to the theoretical notion we resolve to produce *real output* too by deflating the *nominal*, removing the influence of prices from the change in 'quantities' – another abstraction, as quantities are hardly measured. With prices in the picture, a lot more data are now needed – in the limit, the prices for each good and service produced, and for every period.

- The theoretical notion calls for *unduplicated* output. Adding up the value of the wheat produced by the farmer, the flour by the miller, and the bread by the baker will count multiple times the same quantity as the output of one becomes an input for the other. This gives rise to the concept of *value added*. But we can't just go out and ask businesses what their value added is. They don't quite care. They care about revenues, expenses, profits, market shares, and the like. Frequently, even accountants have a hard time with the notion of value added – something that was particularly pronounced prior to the introduction of value-added taxes. But the notion is real and can be constructed, if enough data were collected. So, more detailed measurements are needed for intermediate inputs this time.

- We realize that the value of all goods and services produced is exhausted by its distribution to the factors of production (something tautological when profits are treated as a residual). Then, we can measure output through the income side too. Even more data are needed for wages, rents, interest, and profits.

- We also realize that the expenditure side is even more convenient and, following Keynes and the aggregate demand, of more interest to policymaking. An elegant representation that captures the value of unduplicated output is the famous C + I + G + X − M (consumption + investment + government spending + exports − imports). All output produced, adjusted for net exports, is either consumed by households, invested by businesses, or consumed and invested by the government. In the spirit of kicking the can again, each of these terms now needs to be defined and measured. At the end, we produce all three variants of output. Having declared beforehand that, theoretically, they must amount to the same thing we call the difference that surely arises in practice *statistical discrepancy*. Split it equally, and we're done.

- Then we must deal with the same old issue, country vs. nation. This leads to slightly different measures. Today, most countries use the GDP as their main output indicator but not that long ago it was the gross national product (GNP). The former measures the value of goods and services produced within the borders of a country, the latter adjusts for net investment income between residents and foreigners.

- The biggest matter of all is that, strictly speaking, neither of them is what the theoretical notion of output calls for. Ideally, it should be net domestic product (NDP) or net national product (NNP) by measuring *net* rather than *gross* investment. Part of the investment during a period will simply replenish the capital stock (and inventories) used up in the production process and should be subtracted − if only we had more faith in the depreciation estimates. But such estimates also exist for the connoisseurs.‡‡

This is barely even a synopsis of what's involved. The bottom line is that we resort to gross investment and estimate a *quantity notion* with *monetary units*, while differentiating between GDP and GNP, current and constant dollars, market prices or factor cost, introduce chain linking for rebasing,

---

‡‡ Different assets have different useful lives. Also, estimates are produced by several depreciation methods and techniques, including arithmetic, geometric, and harmonic.

and so much more. Not only we end up with a long menu of variables and components again, for which data are needed, but we generate a huge body of knowledge in the process. If detailed plans were needed to provide a strong underpinning for the creation of the new theoretical structure, we can claim success! A bulky and oversized UN manual on national accounting is dedicated entirely to that.

Equipped with the blueprints, the actual construction can now begin in earnest (block C in Figure 4.2). Like the machinery and the material that will arrive on site for a physical structure, they'll come from many different sources, will be collected through different methods, and their integration will require skills and expertise.

A quick look at the expenditure side GDP is instructive. How do we measure *consumption* for a period? While straightforward for tomatoes, it's not so for the purchase of an automobile. Services derived from *durables* will be consumed over several years (another dreadful depreciation exercise). It's easier to measure *consumer spending*, deviating from the theoretically correct notion. This can be done through a retail trade survey or, going forward, scanner data. Investment in residential and non-residential construction, and machinery and equipment, will need a cross-economy capital expenditure survey. Government spending will need to reconcile all flows between national, regional, and local levels. Even better if the measurements adhere to the international standards for government financial statistics. Merchandise exports and imports can come from customs records, those for services from surveys and other sources. And, incidentally, kudos to the cooperation between Canada and the US that for three decades now use each other's imports as their exports. Not only it saves an embarrassment but it helps researchers who up to that point could not reconcile why Canadian data on exports to the US bore no resemblance to US data on imports from Canada.

National accounting can absorb an awful lot of data, everything we touched upon and then some. This particular exercise is a bottomless pit, and it can take all you can throw at it. As if the integration of massive amounts of real data is not enough, the blueprints also prescribe a large number of imaginative adjustments out of deference to the spirit of the theoretical notion – which has been violated several times already. One of my favourites has been the *imputed rent for owner-occupied dwellings,* which makes up and assigns rent income to homeowners ensuring that the GDP is invariant to the distribution of homeowners and renters.

At the end of a long road, the structure is complete. All estimates are constructed. Now, the data work can move on to analysis and dissemination (blocks D and E in Figure 4.2). Analysis is a separate, value-added activity that pumps up the adrenaline of its believers and makes them forget all the pains associated with the production of the data. Think of it as the actual exploration of the new construct, a discovery tour not only through the lobby and the high-traffic areas but through basement corridors, stockrooms, shafts, furnace rooms, hidden niches – everything. With the blueprints at hand such exploration is well-guided. Early analysis can be exploratory and descriptive in nature, honing familiarity with the data. Deeper analysis is always inferential, causal, and explanatory and goes beyond the immediate data. It requires knowledge of the literature, policy and regulatory environments, the inner workings of actual markets, and linkages with other areas. A good analyst can look at any data set and do a decent job on the relative importance of key components and their movement over time. But inadequate subject matter knowledge invites overt and covert misses, something troublesome as they limit explanations. Subject matter expertise is key to turning data into information, insights, and knowledge. Moreover, analysis feeds back to theory, and even more to blueprints and measurements.

Analysis can also enhance storytelling in dissemination, which can include revealing narratives and appealing visualizations or sound bites from the data and the findings. These can be likened to the finishing touches that contribute to the aesthetics of the new structure. They can be personalized to individual tastes and preferences and be very creative activities, imagination is the limit. While they don't affect the structural engineering of the construction, they do matter in the sale price.

## Overlaps and Implications

The stepwise process described above is common to any notion we set out to measure. While the blocks shown in Figure 4.2 are visually easy to follow, you can just as easily imagine them overlapping. You can even imagine superimposing separate such figures for multiple individual notions on top of each other. The relative size of the blocks will be different, as will the speed of the underlying activities. In the limit, they all become one big thing really, challenging the separation of the beginning from the end. Such a perspective reveals several observations that help solidify our understanding of real-world data work.

*Common data and multiple uses:* The process of scoping, defining, and refining proxy variables to estimate abstract notions invariably leads to the need for additional, and increasingly more detailed, data. This was shown in the examples of both the unemployment rate and the GDP. Unless a particular notion requires a very particular data set, as measurement work is carried out for different notions it shouldn't be surprising that some of the data overlap. The requirements may not be identical but they belong to the same family of data. For instance, income-based GDP requires data on labour income, poverty measures require family incomes from all sources, including labour income; a lot of the price data for constant-dollar GDP are the same as those for the consumer price index (CPI), which primarily supports monetary policy and cost-of-living calculations used in collective agreements, rent increases, and indexation of pensions.

*Data yards:* Understanding how data are actually produced is key. Rather than catering to the needs of an individual notion, setting up independent measurement programs for certain areas has advantages. Regular and ongoing production generates economies of scope and lowers the marginal cost of extra information requirements. Whether through a survey, an administrative or other source, the statistical infrastructure can accommodate the needs of multiple users, up to a point, through flexibility in the determination of content. Several such programs have been set up in statistical offices over time. So, practically, when a new notion emerges there's no need to start from scratch. Using our building metaphor, when it's time for the actual construction of a new structure, yards with stockpiles of materials already exist. Of course, specialized items may be needed. Those working in measurements will recognize that this has implications for all their routines, including IT systems and processing environments, something that shouldn't be underestimated.

*Evolution:* As everything evolves, not only new notions requiring new data appear but rethinking of older notions can lead to modifications of existing data. Decades after the post-war era creation of the GDP, there was a change of mind on some matters. Opinions converged that R&D should be treated as capital rather than a current expense. Same for software, which didn't exist back then. Current thinking is that *data* is such an asset too, and should be capitalized – especially databases. Any such change in notions kickstarts changes elsewhere. For instance, refining the notion of GDP necessitates revisions to its blueprints (e.g., the UN manual), adjustments to feeder sources, and backcasting of historical estimates.

Static data can't serve evolving needs (duh)! When *innovation* came to prominence, it was measured through business surveys. It was determined that the measured proxies would be *product and process innovations*, which can then be tabulated by industry and size of business. Such innovations could be first-in-the-industry, the country, or the world. A whole manual was produced to guide measurements in the new area, the Oslo manual.[4] With experience and fresh thinking, *organizational and marketing innovations* were later added to the mix. The *e-commerce* notion emerged in the late 1990s, when barely a few books were sold online. Speedy work declared online browsing out of scope and differentiated between *shopping* and *purchasing*. E-commerce was defined as the value of sales generated by online clicks, regardless of the means of payment – and in those days cash on delivery was a prevalent method.

*Matters of quality and fitness-of-use:* Data quality is seen as multi-dimensional, encompassing matters of coherence, timeliness, interpretability, and accessibility among others. For our purposes here, *accuracy* is the quality dimension of interest. We'd seen earlier that better instruments can improve the accuracy of physical measurements. But how do we assess the accuracy of measuring abstract notions? The 'instrument' here is really a *method* relying on our choices for proxy variables and input data. The GDP is our current proxy for aggregate output, and a labour force survey our choice of data source for the unemployment rate. Consistent with the exposition all along, our starting point ought to be how closely our estimated variables approximate their theoretical notions. This task can only be accomplished tacitly, yet it's still more important than comparatively lesser accuracy issues. If we all agree that *years of schooling* are poor proxies for *literacy* and *numeracy*, there's no point looking for better data for schooling. In such cases, our ability to compare alternatives can serve as an interim guide. If the CPI is not a good proxy for our notion of *inflation*, what is?

We saw, step by step, how socio-economic data are constructed. Deep drilling is involved, corners are cut, numerous data series from different sources and reference periods are used, discontinuities occur, methodological changes are introduced, and so much more. To pretend that our measures are infallible would be disingenuous. Which is why I much prefer the term *estimates* rather than measurements – it signifies a distance from the absolute truth. Even so, you'll frequently come across the rather oxymoronic 'accuracy of the estimates', and you'll be left with the distinct impression that it can be calibrated through some dial-like approach. This is directly influenced by the trade-off between lower margins of errors and bigger overall effort (bigger sample, more time, higher cost) in sample surveys, and comes with connotations for how 'fit' the data are. It's quite true that if data estimates with a ±5% margin of error are deemed adequate for some intended use, it'd be hard to justify going after estimates with a ±2% margin that may take twice as long and cost three times as much. However, since data accuracy has no intrinsic value unrelated to the intended use, this kind of accuracy is really subsumed by a *fitness-of-use* criterion that first and foremost attempts to ascertain the distance between the estimated variables and their abstract notions. As we discussed explicitly, the most accurate rate of unemployment cannot substitute for the study of underemployment.

*Revisions:* Nonetheless, a common way used to gauge data accuracy is keeping tabs on their revisions. These can be ongoing, periodic, or impromptu. On the one hand, the very existence of revisions is an admittance that the data released are the best available *estimates* of an *unknown truth* and demonstrate transparency on behalf of the data producer. On the flip side, revisions don't really represent a promotional or bragging opportunity. Whichever way you cut it, they put a dent on any perceived precision of the data. This may rub off to other numbers produced by the same source. Thus, data producers strive to strike a balance between up-front transparency and credibility. To remove the element of surprise, whenever possible, revision policies are produced for influential statistics and communicated publicly with pre-announced schedules.[§§]

Revisions incorporate the most current information and/or implement improved estimation methods, such as updates to benchmarks, projectors, and seasonal adjustment. Such work becomes a matter of routine and is baked in the ongoing production of the data. The size of the revisions over time can be instructive, perhaps best seen under the pressure of fast-produced, low-frequency data (e.g., monthly) that are known to be 'noisy' and even prone to erratic swings. Similar is the case of *nowcasting* data, thanks to demands for ever more timely information in our impatient times – to 'forecast the present' and 'estimate the estimates'. Low-frequency and nowcast data are useful as short-term signals of impending change but not the best for accuracy. An analogous argument is frequently made for the GDP. The accuracy of its level isn't really important, what matters for policy is the accuracy of the change from period to period. Therefore, the consistent application of the same methods should inspire confidence. Unfortunately, it's the kind of indicator in which change is measured by decimals.

The epitome of transparency is the recent appearance of *real-time tables* with vintage estimates that show all revisions of a specific data point over time.[5] Then, in the national accounts there are also periodic *historical revisions*, which are substantial as they incorporate changes in scope and concepts, and typically lead to *backcasting* previous estimates over decades. Effectively, such revisions recreate what we thought was the truth! You can appreciate why this gives rise to comments such as *"an uncertain past"*.[6] (Once again, nowcasting and backcasting add perspective to the joke about the meteorologists' erroneous forecasts of the future.)

Useful clues on the impact of revisions can perhaps be had from the intended use of the estimates again. At times, a small change in direction can be more controversial than a larger revision in magnitude. I've seen a modest uproar over revised GDP figures, say from 0.1% to −0.1%, as the direction differs and affects the conventional definition of a *recession* (two consecutive quarters of negative GDP growth). But I can hardly recollect an example when a change of 0.1% in GDP led to an action radically different from a change of 0.3% – if it led to an action at all. More than achieving some ever-elusive measure of accuracy, revisions matter more for credibility reasons. Inaccuracies, under best efforts and good intent, pale in comparison with mischief. Proof comes when things go really awry. There have been enough examples of countries fudging the books, with statistical offices falling prey to political pressures.

Revisions do not typically apply to some data, such as the CPI. Monthly prices, once measured, are not subject to change. Still, much like the other

---

[§§] To illustrate, in Canada the quarterly GDP estimates are revised in each subsequent quarter of the year, and once a year all quarters are revised back three years. In the US, the Bureau of Economic Activity publishes an *advance release* of the quarterly GDP four weeks after the end of the reference quarter and a *final release* three months after the end of the quarter.

measurements we discussed, all sorts of practical matters can affect the theoretical purity of the price notion and the aggregation across multiple prices. From changes to the basket of goods and services to the introduction of new products to outlet bias and a whole bunch more. Arguably, accuracy is more important for the CPI given where and how it's used. Its calculation has even been scrutinized in congressional hearings in the US finding a bias of a few decimals. As we saw in other notions, other uses care about different components, such as the *core CPI* used by central banks, excluding food and energy, or some other aggregate. Even a CPI for seniors has been created.

*The players:* The workload associated with the activities described in this chapter requires many talents. They're mostly found in different people but at times some may coexist in one person. In the extreme, a single researcher carries out the whole exercise single-handedly, from the conceptual development and the design of the blueprints to field work, data capture, processing, analysis, and dissemination – visual effects included. In bigger statistical operations, tasks are performed in different areas and certain individuals have expertise in or responsibility for one or more of the activities involved. Generally, in describing those involved we use terms like academic, theorist, modeller, statistician, analyst, policy expert, and many more. The sure thing is that someone can make a career in any of the underlying activities.

## Fact-Checking Tips

*The fallacy of the broken glass:* In a 1921 film, *The Kid*, Charlie Chaplin played the Tramp character, a glass repairer who, by lucky coincidence for the shop owners, is just a few steps behind a kid throwing stones and breaking their windows. As simple as that, not a more elaborate scheme. This links back to an 1850 tale by Frederic Bastiat *What we see and what we don't see.*[7] A boy breaks a shop window but those around consider it a good thing. The logic? His father will pay the glass guy for the repair who, in turn, will spend it on something else starting the classic chain reaction that we call economic activity. All good – until the author pointed out what was missing from this go-ahead-and-break-things approach. It starts with something that decades later would become famous as the economists' *opportunity cost*. The father had alternative uses for that money; he could have spent it on something else. The next best alternative would have stimulated more productive activities elsewhere than replenishing the broken window. Using resources and expending efforts to fix something that was already there is inferior to building something new. It's like running to stay put. Destruction doesn't pay…

Fast forward to our times. All too often, you'll see severe criticism laid against the GDP for failing to capture the devastation caused by hurricanes, earthquakes, wildfires, and other natural disasters. Numerous studies consistently show a tiny effect that doesn't jive with the images and personal pain you see on TV. For instance, the damage from hurricane Katrina was estimated at US$161 billion, the highest of all hurricanes on record. In addition to the tragedy of the death toll, governments and insurers had to confront the harsh financial realities related to the destroyed infrastructure and the loss of property, vehicles, machinery and equipment. Businesses and individuals came face to face with financial ruin too. Yet, the GDP was more cold than sympathetic to the plight. The Congressional Budget Office estimated that the combined effects of both hurricanes Katrina and Rita in 2005 led to a 0.5% reduction in real GDP growth in each of the third and fourth quarters of that year. It also estimated that a 0.5% additional GDP growth would be stimulated in the first half of 2006, thanks to the enhanced economic activity that would follow.[8] A practical explanation comes from the fact that, outside of small island states where a natural disaster can set them back for decades, in large economies the effects of such disasters are absorbed locally. For instance, the GDP of Louisiana and the surrounding areas affected by Katrina was estimated at around 1.5%–2% of GDP. A more to-the-point explanation is that the GDP doesn't capture the loss of assets, such as roads, bridges, property, or vehicles. Its contraction comes from lower economic flows, such as business closures, loss of employment, and reduction in produced goods and services. In time, resources will pour in and spur activities that may well grow the GDP. Negative happenings can eventually lead to a higher GDP. A burnt forest subtracts from logging, which can be more than compensated by proportionately larger reforestation efforts soon after. Higher crime can increase the GDP, if it leads to higher spending for security and policing or even for processing the extra crime itself. All too frequently the GDP takes the hit for such 'upside-down' measures. But this is not a defect of the particular statistic. Its 'field of vision' has been designed with the ability to 'see' many things and 'not to see' others. One can find plenty of imperfections and flaws and stick it to the GDP, but not this. (Incidentally, it's also fair to compare its caveats against the next best alternative – like we do for all the imperfections and flaws of applied democracy.)

The implication is that fact-checking itself cannot be blindfolded by innumeracy. There is a responsibility to know the true attributes of an indicator, its qualities and limitations, not vague half-truths or heuristics related to what someone thought it could, should, or ought to be. You can't accuse the metre for not measuring volume. From an estimate of the value of goods and services produced, do you stretch the GDP as a proxy for the standard of living? Why, and how far do you go? The GDP definitely never set out to measure a country's wealth or environmental degradation. There are wealth accounts to record the loss of capital assets and deforestation. The GDP doesn't measure well-being, and it doesn't deliver justice. And, no, the GDP doesn't measure happiness!

*Overreaching?* However, the king of Bhutan did. In 2008, the Gross National Happiness (GNH) was embedded in the constitution of the mountainous country with the intent to get past the GDP and onto more of the things that really matter. It represents a holistic approach to development, incorporating non-economic aspects of well-being. No lesser figure than the prime minister chairs the GNH commission. Around the same time, the then-president of France appointed Joseph Stiglitz, Amartya Sen, and Jean Paul Fitoussi to lead a similar effort: identify existing limits and figure out how to move *beyond GDP* and into the nation's well-being through additional indicators of social progress. Such work was picked up by several other countries and the OECD, and there are many initiatives underway by now. Rather than complaining about what existing measures don't or can't do, this is the way to go. Construct new abstract notions, whose subsequent quantification will prove useful to our desired path forward. Can we really do such things? Is it an overreach? Pie in the sky?

The 1776 Declaration of Independence in the US refers to "*Life, liberty, and the pursuit of happiness*" as three of the unalienable rights given to all humans by the creator, but doesn't prescribe measurements. How to measure happiness? Some thinking probably flirted with a temperature-like measurement. We *feel* happier and sadder, as we feel colder and hotter. Can't we just agree on a zero and a unit and off we go? Attempts to create such a unit were made in the context of *utility*, a somewhat related notion that preoccupied economists in the late 19th century. Utility is not happiness, more like the satisfaction derived from consumption. A *util* could be a unit to measure cardinal utility. Francis Edgeworth even proposed a hedonometer! "*...let there be granted to the science of pleasure what is granted to the science of energy; to imagine an ideally perfect instrument, a psychophysical machine, continually registering the height of pleasure experienced by an individual...*"[9] Neither the util nor the hedonometer went far, and by the 1930s such attempts had been abandoned. However, the ordinal utility approach carried on and led to the theory of diminishing marginal utility, under which consumption of additional units adds progressively less to total utility. Such thinking led to indifference curves, and from there to the demand function. Today, besides the work on measures of societal well-being, there's also a revival of efforts to measure happiness and related concepts at the individual level. Surveys employing psychometric techniques and other methods are tried. There's even a modern-day digital hedonometer![10]

However, no widely accepted measure has surfaced yet. It's not merely the absence of an agreed-upon convention that separates us from such metrics. We must contend with a whole host of interrelated notions, wrapped into knots by the confusing signals emanating from our wiring – our human nature. As one example, in *Thinking Fast and Slow*, Daniel Kahneman draws the distinction between *happiness* and *contentment*.[11] Looking at the totality of one's life, say, a family, a job, perhaps a house, accomplishments, and experiences lived, someone may feel content with life in general – but miserable and seeing black right now due to crises in the last few days, to the point of scoring at the bottom of a happiness scale from 1 to 10. (Analogous and purely subjective scales are used by doctors to gauge the intensity of pain.) Content and unhappy at once brings to mind the distinction between stocks and flows again – not a good month for income but the bank account is fine. Think also of Epicurus and his followers in ancient Greece – and today. Life's intrinsic goal is still *pleasure*, but not understood as extreme exhilaration or bliss. Instead, the quest for pleasure is one for tranquillity, with no bodily pain, the absence of fear and, above all, peace of mind. These will come from a simple life rather than material over-indulgences. Then, in our times, some happiness signals register fast and vanish faster, thanks to consumerism and instant gratification. Could *nirvana* become the highest possible point of a happiness scale, equivalent to Kelvin's absolute zero?

A couple of remarks are of practical relevance here. First, the need to show leniency to new efforts. We can't know immediately and with certainty what'll catch, what'll become a breakthrough, or what is utter nonsense and waste of time. Attempts at something new or daring may come and succeed or fail, may fail and never come back or may return – to succeed or fail again. Keep an open mind. Most of our statistics today are fairly young; they didn't exist much before. While many things around us look as if they came out of early drawings by Da Vinci, stories by Jules Verne, or Star Trek episodes, such sources contain more things that aren't among us – yet! Without reneging on being critical when fact-checking, follow what good teachers or coaches must do when confronted with uncertain decisions of pass or fail. A marginal pass needs to be based on signs of future promise, a fail is a better service when it saves someone's time and urges them to find their true calling. The point remains that it's easier to tear a new effort to pieces but more challenging to offer constructive criticism. Give some credit to well-meaning intentions.

Second, it may be true that within our time horizon we may have tricked ourselves into overreaching. There are plenty of examples that we can't handle properly some of our abstractions. We use inputs as proxies for many outputs. We can't quite measure health outcomes but we measure in detail hospital beds, medical staff, waiting times, and cataract procedures. We can't even measure the government sector properly, which accounts for 40%–60% of GDP in some western countries – measured as salaries for civil servants among other expenditures. Other times, our attempts work up to some point before they break down. Income-based GDP treats net interest outlays as a factor payment, representing value added by the paying industry. Tell this to the banking sector (all those taking deposits and giving loans, living off the interest spread), who ended up with negative value added – and they're still exempt from value-added taxes for that very reason. But then again, corrective actions do occur. Someone produced a thesis and solved that too. The game is on.

*Hidden traps:* There's more to fact-checking than the accuracy of data and their appropriate use. From inception to dissemination, the lifecycle of data is rife with opportunities for errors but also for mischief or deceit. Assessing factual information extends to the identification of factors that lurk in the background. Among them are fraudulent data, data intended to seed doubt rather than inform, ulterior motives behind findings allegedly supported by the data, and spin. Tim Harford in *The Data Detective* cites the case of Darrell Huff, a freelance journalist, who'd gained notoriety thanks to his 1954 book *How to Lie with Statistics*. In a hearing at a US Senate Committee in 1965, he applied some of the very tricks he'd previously exposed to concur that the correlation seen in the data between smoking and disease was equivalent to that between storks and delivering babies. It was soon revealed that he was bought off by the tobacco lobby.[12] A common technique against overwhelming statistical evidence is not an outright refutation but attempts to muddy the waters, *seed doubt*. This has happened repeatedly with matters related to climate change, and more recently with COVID-19. It's also known that, psychologically, a shred of doubt provides enough cover for those predisposed to cling on to a lie.

There's a plethora of examples of *fraudulent data* and unethical behaviour in research. Too many documented cases have exposed lying in grant applications, misconduct in actual trials, faking data, falsifying findings, even inventing co-authors. This has been particularly prominent in medical and clinical research but socio-economic research is not immune from analogous shenanigans. Other than penalizing and disgracing those caught, issuing apologies, and retracting publications, there are now all sorts of due-diligence guidelines for research ethics, conflict of interest, requirements for disclosure and the like.

There are also *ulterior motives*. There's nothing wrong for a company to fund research aimed at reversing baldness – many people would be appreciative if they succeed, as many others have been with breakthrough medications. This is very different from well-orchestrated plans by entities to drive nefarious research agendas and recruit or influence researchers to do their bidding by selectively picking or concealing data at will. Beware of studies backed by food or beverage companies meant to shift away the blame from obesity or associations for dried goods that promote the positive influence of cashews on semen.

Lastly, fact-checking always has to put up with *spin* too. This can range from the dangerous to the innocuous – it can even be funny. The implications are different. Catchy headlines to entice you deeper into a story can come even from reputable sources, as everyone is vying for some attention. They're recommended for communication purposes, to convey the 'essence' of findings and free the messages from 'statistical jargon'. In such cases, you'd fully expect the rest of the story to corroborate the headline. An 'overwhelming' majority prefer X can't be based on 43% vs. 36%, with 21%

undecided. Spin can be quite obvious, somewhat subtle, or a full camouflage for things meant to be kept out of sight. For something like *Pasta from Barley Makes you Healthier*, which made the rounds a few years ago, is the headline linked to the conclusions or the whole thing is better explained by other information scattered somewhere – such as the pasta used in the study was 'kindly provided' by a specific maker, together with a research grant?[13] The fine print matters. When the jargon is too discombobulated for mere laypeople, a judgement call is needed. Either way, it may become tough to decipher between a small exaggeration and a big, blatant lie. But when the going gets tough, the fact-checkers get going.

## Notes

1 David Card, "Origins of the Unemployment Rate: The Lasting Legacy of Measurement without Theory," UC Berkeley and National Bureau of Economic Research, February 2011, https://davidcard.berkeley.edu/papers/origins-of-unemployment.pdf

2 US Bureau of Labor Statistics, *Economic News Release, Table A-15, Household Data, Alternative measures of labor underutilization*, Last Modified Date: May 6, 2022, www.bls.gov/news.release/empsit.t15.htm

3 Statistics Canada, *Guide to the Labour Force Survey 2020: Section 2: Determining Labour Force Status*, 7 (Cat. no. 71-543-G), Release date: April 9, 2020, www150.statcan.gc.ca/n1/en/pub/71-543-g/71-543-g2020001-eng.pdf?st=kt2qpNf_

4 OECD. *Oslo Manual*. www.oecd.org/science/inno/2367614.pdf

5 Lately, Statistics Canada started showing entire streaks of revisions in some data releases. Statistics Canada, *Real-Time Data Tables*, Last Modified Date: April 21, 2022, www.statcan.gc.ca/en/dai/btd/rct

6 Term used in a Bank of Canada paper by Greg Tkacz, "An Uncertain Past: Data Revisions and Monetary Policy in Canada," *Bank of Canada Review*, Spring 2010, www.bankofcanada.ca/wp-content/uploads/2010/06/tkacz.pdf

7 The original title in French was "Ce qu'on voit et ce qu'on ne voit pas." See Andrew Beattie, "The Broken Window Fallacy," *Investopedia*, Article updated, July 30, 2021, www.investopedia.com/ask/answers/08/broken-window-fallacy.asp

8 US Department of Commerce, "The Gulf Coast: Economic Impact & Recovery One Year After the Hurricanes," October 2006, www.commerce.gov/sites/default/files/migrated/reports/oct2006.pdf and Ryan Sweet, "How Natural Disasters Affect U.S. GDP," *Moody's Analytics*, 2017, www.economy.com/economicview/analysis/296804/How-Natural-Disasters-Affect-US-GDP.

9 Quoted in David Colander, "Edgeworth's Hedonimeter and the Quest to Measure Utility," *Journal of Economic Perspectives* 21, 2 (2007), pp. 215–225, https://sandcat.middlebury.edu/econ/repec/mdl/ancoec/0723.pdf

10 Under the motto that *"You can't improve or understand what you can't measure"*, a website has *"created an instrument that measures the happiness of large populations in near real time"*. For now, the data source is Twitter! "Happiness," Vermont Complex Systems Center, https://hedonometer.org/about.html

11  Daniel Kahneman. *Thinking, Fast and Slow*, Anchor Canada, 2013.

12  Tim Harford. *The Data Detective: Ten Easy Rules to Make Sense of Statistics*, London: Riverhead Books, pp. 14–17, 2021.

13  You should also know that this particular study involved feeding 30 adult mice for five weeks with a low-fat diet, half supplemented with BBG-enriched pasta (barley), the other half with regular wheat pasta. Quoting verbatim from the conclusion of the study: *"To the best of our knowledge, this is the first study to show in vivo data demonstrating that a sustained dietary intake of pasta enriched with BBG safely increases coronary collaterals, limits the infarct size, and reduces mortality. This is achieved through the simultaneous myocardial upregulation of p53, VEGF and Parkin without interfering with the ability of cardiac cells to phosphorylate Akt, STAT3 and NF-kB during the post-ischemic reperfusion. Higher p53, VEGF and Parkin protein levels in endothelial cells chronically treated with BBG are not affected by exposure to acute oxidative stress. Finally, Parkin responsiveness to BBG is not driven by the inhibition of p53 activity, which in turn may contribute to VEGF expression and anion super-oxide decay"*. See Study by Valentina Casieri, Marco Matteucci, Claudia Cavallini, Milena Torti, Michele Torelli, and Vincenzo Lionett. "Long-term Intake of Pasta Containing Barley (1–3)Beta-D-Glucan Increases Neovascularization-mediated Cardioprotection through Endothelial Upregulation of Vascular Endothelial Growth Factor and Parkin." *Scientific Reports* 7, 13424 (2017), https://doi.org/10.1038/s41598-017-13949-1

# 5

## The Art of Drawing Lines

It's now time for an inside look at *how* statistics are produced. We'll also get to meet some key tools of the trade up close, and understand how they're constructed and used. Tracing the making of statistics back to their beginnings will shed light on the thinking that takes place and the decisions that have to be made along the way. This process of discovery will reveal their true meaning and demonstrate their strengths and limitations. All that becomes critical later, when the time comes to derive insights and knowledge through data interpretation and analysis.

### The Opening Act

*"This is so unfair. It doesn't make any sense. It's killing my business. What are those people in city hall smoking?"* These are the words of a pub owner, uttered angrily in a microphone extended to him on a long pole by a reporter while the camera is filming. The segment will air on the local television channel later tonight. Several bystanders have gathered around, attracted by the commotion and curious to see what the fuss is about. The camera focuses on the pub's patio, before it rotates across a wide two-way road to film a similar patio of another pub on the other side.

Earlier in the day, the city passed a new by-law prohibiting smoking even on outdoor patios. The irate owner will have to abide, but the pub across the street won't be affected. The street that separates them is also the street that separates the two municipalities to which they belong. Situations like this played out in real life several times and in several places over the last couple of decades. Leaving aside the desirability of having adjacent municipalities in urban areas coordinate in such matters, or arguments to the effect that the complaining pub owner may actually gain from non-smokers who'd flock to his side, at first glance some may sympathize with his plight. The optics are really terrible. Any passer-by with half a wit would have never thought of imposing different rules on those two pubs. Looking around, they're just a few metres apart and there's nothing much separating them except that they're on opposite sides of the same street.

DOI: 10.1201/9781003330806-5

The fact that the middle of the street is also the border between two municipalities is nothing to raise eyebrows, it's quite common. Typically, cities have a downtown core, surrounded by parts in the middle, which extend further out to the periphery. At some point, city limits are demarcated by an outer perimeter. Exactly the same mapping procedure is used to separate neighbourhoods within cities, albeit not for jurisdictional purposes. The fundamentals of such line drawing are not even affected by amalgamations in big urban areas. However, the issue of how perimeters are delineated is a legitimate one to look at.

## Know No Boundaries?

There's line-drawing galore everywhere in our world, and for all kinds of reasons. Demarcation of sovereign national borders is known to lead to disputes, even wars. Inside countries, divisions of regional and local areas extend all the way down to postal or zip codes, even individual house lots. There are even lines on maps that show the zones for local pizza delivery.

Historically, the morphology of the land itself has provided plenty of handy dividers. Mountain ranges, cliffs, rivers and streams, forests, and other physical barriers become natural separators. Train tracks, highways, and other man-made structures also provide convenient references. Frequently, though, arbitrary lines too need to be drawn. Sometimes they're kept remarkably simple and straight. Just look at the maps of Colorado and Wyoming in the US or Saskatchewan in Canada. The same simplicity can cause the occasional confusion. A good example is the border between Canada and the US, the longest in the world at 8,891 km – counting Alaska. Drawn as a straight line on the 49th parallel from the Pacific ocean to Manitoba, it goes through the Great Lakes dipping all the way south to lake Erie, cuts right through Niagara Falls as it starts to move north again, and follows the middle of the St. Lawrence river for a while before it veers further north and becomes squigglier all the way east to the Atlantic ocean. From all that, it's the 2,050 km long (theoretically) straight line that creates the mental image that Canada sits on top of the US. Friends in Seattle had to turn to Google in disbelief that they are north of Toronto, which at a latitude of 43.7° N is four degrees lower and at the height of Marseilles on the Mediterranean coast.

Outcomes from drawing lines are oftentimes constructive. Good fences make for good neighbours, and clear communication makes for good friends. Lines help the arrangement of many of our affairs, such as the area of jurisdiction for various levels of government, the use of land through zoning designations, and the establishment of electoral districts. Sometimes, though, such exercises can turn more sinister. Extreme gerrymandering in the US for electoral redistricting is a well-known example of how opportunistic cherry-picking of individual neighbourhoods can lead to weird configurations. Other times, different anomalies occur. The Canada–US border goes straight through a town, splitting it into Derby Line in Vermont and Stanstead in Quebec. The library itself was deliberately situated *on* the border, with black masking tape used as a separator inside the building. The entrance is in the US, the reading room in Canada.

Except for actual mapping, we marked the globe with imaginary meridians and parallels defining exact coordinates for real spaces. With GPS and a smart-phone app, no more *"left at the gas station, straight until the brown house, second right after the church"*. How neat is that! We stepped on it to decide matters of time too. Parallel meridians, 15° apart, define 24 time zones. We imagined the *international date line* and conveniently situated it 180° from the first meridian, in the middle of the sparsely populated Pacific Ocean. But we have to con-tend with the issue of the two sides, no escape. Apia, the capital of Samoa, is about 125 km away from Pago Pago, the capital of American Samoa. They're so close that they share the same time. When it's 9:07 am in Apia, it's also 9:07 am in Pago Pago – but on Friday rather than Saturday. Then, once again, we don't adhere strictly to our creations but improvise conveniently. Not only this particular 'line' is extremely crooked, but it also produces counter-intuitive differences of 26 hrs between two places – as that between the Line Islands of Kiribati, and Howland and Baker islands (Figure 5.1). To follow local time conventions in Alaska and Siberia, the islands of Big Diomede (Russia) and Small Diomede (US) are separated by 20 or 21 hrs depending on the time of the year, despite being literally next to each other. All of China uses *Beijing time*, even though the country spans several time zones. But Xinjiang in the west also uses *Urumqi time*, which is 2 hrs behind. Scheduling anything there requires extra communication.

Drawing territorial lines on a map can be quite controversial. Throughout history, and around the world, horror stories abound. Some borders have assumed physical form, with fences, barbed wire, or walls and are guarded 24/7. To this day, notorious and nasty disputes keep many on their toes. Even the Canada–US border took a century of negotiations and compromise. At a detailed level, there are always hidden truths on the ground even for what appear as straight lines on a map. There's always something to go around, spoiling the straightness. Inside each country, the boundaries of geographic divisions are dictated by a combination of historical and practical consid-erations. Moreover, different boundaries are frequently superimposed over the same area to facilitate different intended uses. Either way, very detailed mappings have been created to delineate progressively smaller geographical areas. From provinces or states to counties, cities, municipalities, all the way to neighbourhoods. Less visibly, detailed plans for individual lots are also drawn. In countries with land registers, all pieces of land are measured and recorded. The psychological influence of such areas on our identity is quite instructive, and one of the early topics of conversation among two strangers who meet for the first time. *"Where are you from"* seems to be of interest to everyone.

**FIGURE 5.1**
The international date line

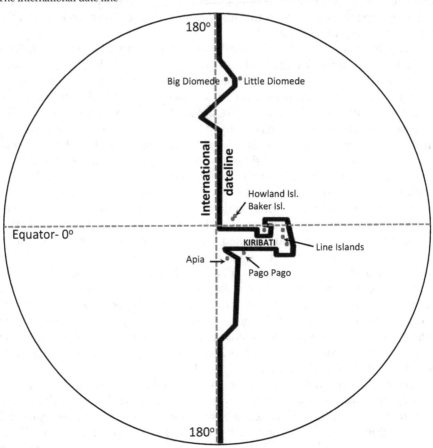

Beyond physical mapping, drawing lines is widespread everywhere as we'll discuss in this chapter. We drew several in Figure 4.1 to define the rate of unemployment. One line cut out the population less than 15 years of age, another excluded the retirees. More lines took care of discouraged workers and other out-of-scope populations until we isolated the particular area desired. As discussed then, what really matters is that we're cognizant of all this.

The decision to draw a line, either because it's necessary or it's deemed useful for some purpose, is distinct from – and matters more than – exactly *where* it'll be drawn. This is so because coming to terms with the decision also means that we're prepared to accept the consequences that will ensue. Surely, we can agonize over exactly where to draw the line. Weighing the pros and cons of different options is worthwhile. But nothing will change the fundamental fact that, by definition, a line will create two sides. Asking why a particular road is the boundary between two areas, as in the case of the pub owner, may be a good question if there's still discretion to make a choice but it ceases to be productive afterwards. It becomes the same as *"why are the 10 commandments 10"*, and a waste of time. Still, I've been a witness to many wasted hours of arguments over exactly such points – and probably you have too. Why is it not a good question? Because it would still stand and be applicable if the commandments were 9, 11, or 12 – or any number for that matter. *"Why this road"* could still be asked if the boundary was shifted to the next road or the one after that. As a rule, any time a question remains intact after a change or some reconfiguration is one more indication that this is not a productive path to follow. Once drawn, someone or something will stand on one side of the line and not the other – and it'd be a matter of perspective if that's left or right, up or down, in or out. We have to understand this for what it is. What's more, we have to ponder its implications – as we'll see later, data analysis has a lot to say on this.

The interest of this chapter in all this is only somewhat related to the fact that the geographic dimension is relevant to the production of statistics. Primarily, it stems from the uncanny parallels between the drawing of lines in the physical world and their equivalent in our conceptual real estate. With seemingly straight lines or zig-zag and squiggly ones, we constantly separate data on sides. If the temper tantrum of the pub owner deserved a segment of television coverage in the nightly news, analogous issues related to our statistics could easily fill an entire channel with round-the-clock broadcasts for a very long time.

## A Sense of Taxonomy

We touched upon matters of homogeneity and groupings in Chapter 1, by imagining our distant ancestor's iffiness with *like* units without the benefit of inherited taxonomies. To this day, every kid – and not only – rediscovers the fun of pondering if the tomato is a fruit or a vegetable. Whether you eat it as a veggie during dinner or with your fruits later is your choice. However, when it comes to accounting for the weight or the cost of your fruits and vegetables, you'll need to ask *"where's the tomato?"* Hidden in the data we produce and use are many answers to *"what side of the street"* and *"where's the tomato"* questions. Sometimes we know them, many times we don't. They have a tremendous effect on every quantity we care about. From the population of Toronto (or Tokyo or Sao Paolo – just try it) to the size of the restaurant business or the employment of plumbers. Obviously, classification decisions that determine the allocation of individual units affect cross-sectional data immediately – by definition they determine the relative shares of constituent parts. They affect time-series data even more due to compositional shifts over time, something more difficult to figure out in practice.

Invariably, classifications employ a hierarchical structure, with broader classes or categories progressively nesting more specific ones. The categories must be unduplicated, mutually exclusive, and exhaustive with regard to the target population. Classifications used to organize socio-economic statistics are commonly referred to as *standard*. This is wishful thinking for the desirability to use them widely, systematically, and consistently across their subject matter. We'll have a good idea of all that by the end of this chapter.

## Geography

The apparatus for organizing statistics along geographic divisions is fairly well developed, nationally and internationally. While ongoing change is inevitable, at any given time we can rely on the latest version of recommended standards that delimit world regions, countries, and other areas. For statistical purposes, the UN M49 standard assigns a three-digit numeric code to every country or area. It's frequently complemented by a three-letter alphabetic code assigned by the International Standard Organization (ISO 3166).* Thus, Canada is CAN 124, the US is USA 840, and Japan is JPN 392. From then on, national *standard geographical classifications* (SGC) delimit in detail distinct areas inside countries. Generally, three levels are common, roughly corresponding to national, regional, and local levels. For instance, in federal countries the second level can be provinces (Canada), states (US and Germany), or cantons (Switzerland). Elsewhere, they can be prefectures (Japan) or something similar. The third level is reserved for lower areas that contain cities, towns, and villages.

In Canada, the 2016 version of the SGC provides names and codes for 10 *provinces*, 3 *territories*, 293 *census divisions* (typically counties or regional equivalents), and 5,162 *census subdivisions* (municipalities, really). A fourth level in its hierarchy consists of six regions to accommodate aggregations of provinces (Maritimes and Prairies). Census tracts are also recognized as smaller and relatively stable geographic areas (usually with 2,500–8,000 people) located in Census Metropolitan Areas. Names and codes are also assigned for additional aggregations of interest, such as census agglomerations, census metropolitan influence zones, economic regions, agricultural regions, and consolidated subdivisions.

In the US, the Census Bureau uses a two-tiered system to delineate subnational areas. Nine *census divisions* are nested within four *census regions*. *Counties* are the primary legal divisions in most *states* and are complemented with *county equivalent* units, such as *parishes* in Louisiana, *boroughs* in Alaska, *independent cities* in a few states, and comparable areas in overseas territories. In 2020, there were 3,143 counties and around 100 comparable areas.

---

* A two-letter alphabetic code is also assigned, such as those used by Internet top-level domain names. Moreover, ISO assigns a subdivision level with two letters preceding the country, for example, ON-CAN for Ontario, FL-US for Florida.

Half the US population lives in less than 5% of the counties. Legally defined subdivisions of counties are *minor civil divisions* (MCDs), commonly known as *towns* or *townships* but can have other designations too. Not all states have MCDs but the five boroughs of New York are recognized as such. There are also *census county divisions*, *incorporated places* that can cross county and MCD boundaries, and *consolidated cities* from the merger of incorporated places and their parent counties or MCDs. Further subdivisions are *census tracts* and *census blocks* which, as the name suggests, are like city blocks in urban areas. Different designations of areas exist for purposes other than mapping, such as school and congressional districts.[1]

Many other countries also have special designations for large geographic areas between the national and the regional levels. Whatever the individual peculiarities of countries, and notwithstanding ongoing changes for administrative reasons, lines have been drawn everywhere. They serve many needs and a whole host of interests. Households, businesses, buildings, and other engineering structures can be mapped and geocoded, enabling the integration of statistics for spatial analysis of economic and social phenomena. Flexibility also exists to cover different needs as they arise by drawing alternative lines at will. For instance, the policy interest in all countries for the different circumstances of urban and rural areas may not be satisfied by designating them based on existing lines. The cores of some small towns may be more densely populated than the periphery of urban centres, complicating some specific policy objective. In such cases, a different look would be warranted – so long as the data can support it. Practically, population statistics for different geographical areas rely on national censuses and, therefore, on the notion of residence. This differs from where people work and play and has given rise to *functional geographies*, which complement administrative geographies by adjusting resident populations with commuter flows in and out of neighbouring areas.[2] They're particularly useful in very populous places – for example, Mexico City, whose metropolitan area includes its municipal namesake and another 60 municipalities. Functional geographies are used chiefly for reasons of employment but people's mobility can be tracked for any other reason too. More recently, data from smartphones were used to gauge mobility patterns of populations under COVID-19 restrictions.

## Grid Systems

Looking from the sky, natural contours are visible. We see oceans, islands, mountains, valleys, lakes, and forests. We also see what we've done to the land. A tapestry of rectangles, other quadrilaterals, and irregular shapes become the prevalent sight from an aeroplane overland. The fictional lines we have drawn to separate countries and regions are mostly invisible but we know they exist in minute detail in our maps. If need be, a good spray on the ground can do wonders for a while. The same is true for our neighbourhoods, our house properties, even the layouts inside our homes. We've mapped everything in excruciating detail, with lines that enclose and separate everything at once.

One would think that we can now lay back and enjoy the fruits of all this labour. But, no, we're not about to relinquish our right to explore alternatives. Perhaps chasing after something simpler or more orderly, we invented grids. The imaginary meridians and parallels offer such an example on a planetary scale. However, we can adjust the scale, we can bring it down a notch or two. With straight lines going down intersecting straight lines going across, we can create boxes of any size we like. Throwing them like nets over the world, or any map, they define alternative territorial divisions. Perhaps they'll be helpful for something.

Several grids have been developed for different purposes. The World Geographic Reference System was created for aircraft navigation but it can be used for global grid mapping too.[3] It can even turn to 3D by adding altitude above the surface. Many countries have defined their own grid systems based on national plane coordinates. Although projections from the globe to 2D surfaces are subject to scale and area distortions, such as the Mercator projection that results in an oversized Greenland, their proponents argue that by omitting direct spatial references grids reduce complexity and help harmonize data sets. Leaving distortions aside, there's a discordance between real areas and their parts that may fall inside cells of a grid – they're unrecognizable. How much so depends on the scale. Superimposing them over our squiggly maps doesn't work. Nothing fits right. Maybe one-quarter of Egypt falls inside a cell. Perhaps a small island country and a piece of Australia fit inside another. Large countries span over several cells. If the cells are small, a city neighbourhood may hardly fit in. Although in any classification there are cases where cutting corners to squeeze things in becomes inevitable, here entire body parts need to be cut out. The shoe simply doesn't fit. However, such grids are totally stretchable, from planetary to tiny scale. Using them to create new imaginary areas, mostly of equal size, can have uses. They can certainly help allocate workload and monitor progress in spatial activities, such as searches. They're used for statistics too.

---

## Making Data

The process of drawing lines to produce detailed and useful maps carries on in the conceptual space. It just employs another nomenclature to accommodate the attributes of the concepts we're interested in. With geography taken care of, we make use of as many equivalents of natural divides as we can find. Gender and the structure of the education system can substitute for rivers and cliffs. We draw the lines at male and female, preschool, high school, and university. Other lines can be subject to negotiations, as has been the case for some physical boundaries. There will be an agreement on an age to separate 'young' from 'old', and whether to draw lines every five or ten years.

## From the People to the Stockroom

Now, let's picture ourselves at the end of a census several decades ago. 'End' means that all the paper forms have been filled and returned. They're all in boxes now, piled up in a huge stockroom. This is *the census*. Each family has responded truthfully to the questions asked of it, and each form contains that family's data – as in *observations*. The truth about a whole country at that point in time is all there. Following months of thinking, preparations, efforts, and expenses, what's been accomplished so far by the whole exercise is to extract data from the people's heads, ink them on paper forms, and collect them in boxes. All the people's microdata are now in the stockroom. No other copies exist, there's no backup. What happens now?

We can't just finger point to our precious collection, claim we have the data, and turn away. As they are, the data are useless. No one knows anything or can do anything, a tangible example of the difference between data as observations and data as information. Using an army of people to count the number of forms in each box and tally them up would get us as far as the number of households. Whatever that total, each form contains information about *all* household members. Do we start all over for the next question that'll surely jump up? What's the population of the country? How many people live in each part of the country? How many men and women? Young and old?

Clearly, this links back to the raison d'être of the whole exercise. We can't arrive at the point of having the data in a stockroom without having thought beforehand what we'll do next. Visualizing the end outputs dictates the earlier steps. Not only the processing of the data contained in the returns but the content of the questionnaire itself should be driven by the end outputs. In a well thought-out exercise, all that would have been planned ahead of time. Remember that for your next project.

## Back to the People

So, how do we produce information and give it back to the people who gave us the data? Rewind to the beginning of the census. Through consultations with stakeholders, and compromises along the way, the output shown in Table 5.1 was designed (imagine the table empty). To arrive there, several decisions were made and lines were drawn. A choice on the detail of geography was made. This would help the initial sorting of returns. Rather than throwing boxes in the stockroom indiscriminately, they'll be labelled by region and

**TABLE 5.1**

Creating and populating a table with data from a census

| Location | Male | Female | <15 | 15–65 | >65 | Feature 1 | Feature 2 | Total |
|---|---|---|---|---|---|---|---|---|
| Region 1 | 245 | 255 | 95 | 320 | 85 | 300 | 200 | 500 |
| Area 1 | 147 | 153 | 65 | 175 | 60 | 200 | 100 | 300 |
| Area 2 | 74 | 76 | 15 | 120 | 15 | 80 | 70 | 150 |
| Area 3 | 24 | 26 | 15 | 25 | 10 | 20 | 30 | 50 |
| Region 2 | 170 | 180 | 80 | 200 | 70 | 230 | 120 | 350 |
| Area 1 | 123 | 127 | 55 | 140 | 55 | 200 | 50 | 250 |
| Area 2 | 47 | 53 | 25 | 60 | 15 | 30 | 70 | 100 |
| Region 3 | 75 | 75 | 15 | 120 | 15 | 90 | 60 | 150 |
| Area 1 | 47 | 43 | 10 | 70 | 10 | 50 | 40 | 90 |
| Area 2 | 28 | 32 | 5 | 50 | 5 | 40 | 20 | 60 |
| Total | 490 | 510 | 190 | 640 | 170 | 620 | 380 | 1,000 |

**TABLE 5.2**

Transcription of a batch of census data to a table

| Region 2, Area 2 (Batch 1) | | | | | | | | |
|---|---|---|---|---|---|---|---|---|
| Id # | Male | Female | <15 | 15–65 | >65 | Feature 1 | Feature 2 | Total |
| 1 | 2 | 3 | 3 | 2 | 0 | 1 | 4 | **5** |
| 2 | 1 | 1 | 0 | 2 | 0 | 1 | 1 | **2** |
| 3 | 0 | 1 | 0 | 0 | 1 | 1 | 0 | **1** |
| 4 | 2 | 1 | 1 | 2 | 0 | 1 | 2 | **3** |
| 5 | 1 | 0 | 0 | 1 | 0 | 0 | 1 | **1** |
| ... | ... | ... | ... | ... | ... | ... | ... | ... |
| 50 | 47 | 53 | 25 | 60 | 15 | 30 | 70 | **100** |

Note:   Bold values in the table match those of area 2, region 2, in Table 5.1.

area as they come in. Sex, age, and other decompositions of interest will be recorded across. So, male, female, and the age groups listed exhaust the total population. The 'feature' columns stand for any number of actual attributes and don't have to exhaust the total. They could represent levels of education, languages spoken, or anything else.

Such prior decisions are reflected in the layout of an empty (Table 5.1) and help guide the design of the next step, *transcription* to ledgers. Someone will be allotted a batch of questionnaires from a certain area (say, 50 returns from area 2 in region 2), with instructions to fill in Table 5.2. Rather than names, assign a sequential number to each return, a record id, and capture the data in each return by filling each row from left to right. Names are the crucial bits of information for genealogy but not relevant for statistical work. This anonymization also protects the privacy of individuals. (Such *coding* can be introduced everywhere else.) So, the first row records a family of five, with a mother and father, two daughters and a son all under 15. The second row records a couple with no kids, the third a widow over 65, followed by a couple with a son, a single adult male and so on. The 50 households have 100 individuals. The subtotals created at the end of each batch will be transferred to another ledger for consolidation until the whole area in each region is covered through a vertical stacking of such sheets. With enough patience, all the forms will be exhausted. The totals will be recorded in the non-shaded cells in Table 5.1. Then, work moves entirely to ledgers. Area data are aggregated to regional and then to national totals. The end results will be produced from the lowest level, one row at a time. Along the way, it will be prudent to double check for human error, and generally implement some quality control procedures.

The essence of the work just described is not conditional on the presence of paper forms. It's equally present in the databases of our digital times, and provides answers to frequent questions concerning the 'mystery' of data processing, including activities such as coding and quality checks. Even in earlier times, there was help from technology along the way in the form of tabulation machines. Moreover, paper-based processing is still the reality in election counts – or recounts. For better or worse, paper ballots are still relied upon to save the day occasionally. Representatives of all parties inspect each ballot before it's placed in the appropriate batch to be counted, precinct by precinct.

Our cherished statistics depend crucially on the multitude of lines we draw. Their totals emerge from the smallest areas or boxes that host the observed data. The same boxes also put a cap on the maximum extent of granularity that can be had. The dependence of the end statistical outputs on these prior decisions places under a new light the earlier assertion that negotiating *where* the lines will be drawn is both legitimate and important. This was particularly the case in earlier, less technologically advanced times. Data producers and users were practically stuck with the choices made upfront, very little flexibility existed. Analytical activities couldn't venture much farther either.

Some variables have been easier to deal with than others. For example, with gender as a binary choice drawing one line was sufficient to separate male from female. Admitting more gender choices from now on will necessitate more lines. Age requires decisions of a different type, with implications on the design of the questionnaire itself. Does a researcher ask for the *age* of each household member or their *dates of birth*? With age, immediate transcription is easier. Dates of birth can lead to more accurate calculations, and they're preferable for later years. On the flip side, they require extra calculations upfront. There may be instances in research when age is captured only in categories. If these are ten-year increments, the flexibility to switch to five-year increments later will not be there.

Categories are also common for income data in surveys, mostly because such questions tend to suffer from low response and lower accuracy. What could provide the conceptual equivalents of natural dividers in such cases? Is it the poverty line or the tax brackets? Almost always round numbers are favoured, more for psychological reasons. Lines can be drawn for equal or unequal groupings, like <$20K, $20–$50K, $50–$100K, and >$100K. Each income grouping will contain a different number of households or individuals. These approaches cannot support alternative statistical outputs and analyses, such as percentiles. Exact income observations are necessary to produce data by quartiles or quintiles, which divide the population in four or five groups of equal size and report the average or median income for each, together with their respective lower and upper boundaries.

In addition to the peculiarities of individual variables, decisions on where to draw lines were historically influenced by the capacities of both data producers and data users. In earlier times, producing data within a reasonable time was more of a factor. Today, electronic capturing and some code help timeliness, and offer more flexibility. The straitjacket of the initial design has been loosened, no need to go back to the stockroom for something missed. For instance, we can use age groups to construct the population pyramids we encountered earlier. Then, we can convert to age in one-year increments constructing even more detailed pyramids. All that agony for where to draw the line feels a bit lighter today. On the use side, the reported outputs had to satisfy user needs, and in a way that the messages could be absorbed by the audience – perhaps with help from the media through 'lift and shift'. Overall numeracy wasn't high and the capabilities of even avid users could only go so far in a world of paper and, later, calculators. Today, there are many sophisticated users, they can manage well on their own. Nonetheless, yesterday or today, it's no surprise that whatever the end product some people would find it excessive, others not enough. This is a predictable outcome, the fate of drawing lines! However, the future points to the direction of data sets rather than tabular outputs – increasingly supporting customization. Undoubtedly, everyone feels a bit more daring in our digital era. Still, the fact that we can do the old things easier and faster doesn't translate to better days at the office. It means that data producers and users are busier, all doing so much more.

## Full Circle

A mere definition involves drawing lines. So do discriminatory pricing practices targeting specific locations, times, or age groups. A promotional offer does the same. Offering a 10% discount to seniors over 60 should not have the 59-year-olds up in arms that they *"missed it by that much"*. We discussed early in this chapter that an explicit recognition of the act of drawing lines and the arbitrary choices involved is required to cope with their consequences. We need to be aware of what's involved in that act, be upfront with it, not fight it or play the ostrich. If anything, we can become better at drawing lines.

Otherwise we run the risk of exposing our unflattering innumeracy side, much like the profound misunderstanding of probabilities when someone feels that having a five in a lottery ticket was *"just one number away"* from the six that was actually drawn.

We must also be clear that despite the astonishing number of lines everywhere, one will be hard-pressed to ever come face to face with a straight one. There's hardly such a thing. Some could appear deceptively straight, and it may be convenient to pretend they are. But there's always something to get in the way of their straightness. Remember our previous discussion about the relationship between the accuracy of measurements and the choice of unit? Let's put it to another use. Say that rather than producing statistics by age group, we go all-out and produce accurate numbers for each and every age. That is, there will be a cell in a table with *Age 10: 175,639*. Users will love it. But what is age 10? It starts with some people who had their tenth birthday on census day – they're counted. It then continues to include many others who will be up to 364 days older (ignoring leap years). A magnifying glass or its equivalent, finer grid lines, will show that even *age 10* is really another group. If we measure in months, age 10 will include months from 120 to 131.

Making the grid finer reveals more detail, such as in Figure 5.2 that shows quite a bit of activity between the cells for ages 10 and 11 rather than emptiness. Is there an end or do we end up in a fractal? If our databases allow – and

**FIGURE 5.2**

Measuring age in months

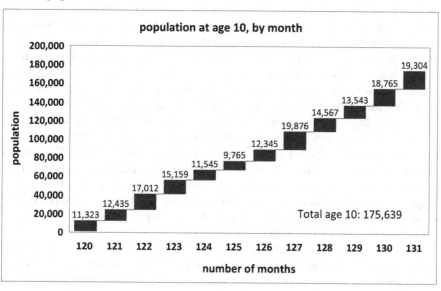

they do if we collect birth dates – we can measure age by days alive, and keep going. What we see as straight lines are the lines of the slot in the grid that houses the data, not the real contour of the microdata. Remember the digital time story? Today we tend to be more precise than in the past. Such thinking also implies that it's possible to do away with some lines and their restrictive consequences rather than trying to straighten them. It's time to stop overrelying on linearity.

## Nothing Personal, All Business

Earlier, we discussed grid systems that divide spaces. They do so neatly, avoiding real-world complications. The conceptual equivalents of grids are classification systems used to organize and present statistical information. Predetermining slots to squeeze in units of a population is what classifications are all about. We'll use one, with application to businesses this time.

The starting point is another stockroom, this time full of boxes from a census of businesses. Or it could be income statements and balance sheets from the tax authority for every unit in the entire *business register*, the list of all active businesses. How do we derive useful information? Randomly browsing through the revenues and profits of a dozen businesses can perhaps yield interesting anecdotes but no coherent information for policy, research, or wider use. We need to recreate a process analogous to that of households. We do have what's needed for a geographical classification of businesses, from their province of operation to their street address and postal code. Sooner or later we'll use that too, but location won't be the starting point. Economists thought that inside a country, with free movement of goods and services, *what businesses do* matters more than *where they are*. Their activity, and more specifically what they produce, is a superior way of grouping businesses rather than their spatial distribution. The notion of *industry* was created. This is not meant in the sense of manufacturing activity, which conjures images of factories and chimney stacks, but in the sense of sectors that include everything produced, tangible or intangible, by for-profit businesses or any other entity. Europeans prefer the term *activity*. So, several decades ago industrial classification structures were created to organize data by grouping *like* businesses together. Industries became names of cells on a grid thrown over the whole economy. Before we tackle this particular conceptual piece, a few things about businesses and products are in order.

Perhaps it's a truism but, as units, businesses differ from individuals. They range from small, home-based ventures to gigantic multinational conglomerates. They can be incorporated or not. Generally, *establishments* are the smallest units with the record-keeping necessary to estimate output, such as employment and other inputs of production. Establishments can operate in one or multiple *locations* but typically in the same industry. Legal entities with varied complexity and ownership structures can form *enterprises,* with establishments across several industries. The head offices of such companies can be good sources for information related to decisions taken at that level. The demographics of businesses are also different. While they too are born and die, they can also morph through such transactions as mergers or acquisitions. Decisions must be made if one ceased to exist or if both did and a new one was born. Ownership, location, and line of business serve as criteria to help determine the continuity or *time travel* of businesses. Businesses can produce one product or a dizzying number of them. Moreover, their activities and output mix change over time.

In principle, classification of businesses is preceded by classification of products. Literally, millions of goods and services exist today and innovation continues to lead to product differentiation. Voluminous directories are dedicated to screw types alone. Classifications need to strike a balance between such unmanageable detail and practicable groupings. An industry classification will typically stop at footwear manufacturing, for example. Sizing the markets for laced leather shoes, loafers, sneakers, slippers, sandals, or flip-flops needs different tools. Some footwear manufacturers will produce one or more of those. Others may have some in their product lines, together with T-shirts, caps, and exercise equipment. So, how are firms actually classified in industries?

The process works along the lines of the example shown in Figure 5.3. Without loss of generality, the value of the goods produced and the revenues

**FIGURE 5.3**
Businesses and products

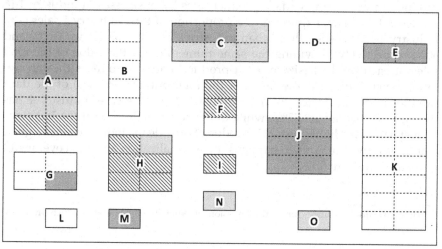

from their sales are collapsed in one. The rectangles indicate businesses of different size, and cells of equal size represent the value of product sales. Starting with a selection of businesses (A–O) and three products (dark, light, and striped), the revenues of each business from each product are shown in matching shades. White serves as a numeraire good; it indicates all other products (not necessarily one white product). Businesses E, I, M, N, and O are 'pure plays', generating 100% of their revenues from one of those three products. Businesses C and H produce two of the three products, while business A produces all three. Business G produces and sells dark too but it's primarily in a different business.

Businesses will be classified in industries based on their *principal activity*, defined by the product that generates the highest proportion of their revenues. Individual businesses will be classified at the lowest level, making it possible to produce statistics for that level but also to roll them up to higher levels. So, A, E, J, and M are placed in industry 1, C, N, and O in industry 2, and F, H, and I in industry 3. The size of both *industries* and *product markets* can now be calculated (Table 5.3). To arrive at industry size, all the revenues of each business will be allocated to the industry of its classification. For the size of product markets, only the revenues associated with each product matter. As can be seen, the two are not the same.[†] Occasionally, this causes unnecessary confusion. The relationship between industry and product markets can be captured with additional metrics, which can also help a deeper understanding of the notion of industry. We have all the data necessary to construct two measures that shed light on the homogeneity of each industry.

The *specialization* of an industry can be calculated by dividing the total product revenue of all businesses in the industry by the total revenue of the same businesses from all products (industry revenues). In other words, how much of the total revenues of industry 1 is generated by sales of the dark product. In our example, dark product sales by businesses in industry 1 (6 + 2 + 6 + 1 = 15) are expressed as a proportion of total industry 1 sales (23). Industry 1 is less specialized than industries 2 and 3 (65.2% vs. 75.0% and 80%, respectively). Dividing the same numerator by the value of the total sales of each product (size of each product market) produces the *coverage* ratio, which indicates how much of the total output (or sales) of the dark product is produced (or sold) by industry 1. In this case, industry 1 has the highest coverage (83.3%), while industry 2 has the lowest (54.5%) since almost half of the light product is produced outside the industry – and a good chunk of it by firm A. It so happens, the specialization and the coverage of industry 3 are both 80%.

---

[†]  They also don't balance due to the omission of other industries and products from our example.

**TABLE 5.3**

Size of industries and product markets

|  | Ind 1 | | Ind 2 | | Ind 3 | |
|---|---|---|---|---|---|---|
| Industries | A | 12 | C | 6 | F | 3 |
|  | E | 2 | N | 1 | H | 6 |
|  | J | 8 | O | 1 | I | 1 |
|  | M | 1 |  |  |  |  |
| Total |  | 23 |  | 8 |  | 10 |
| Product markets | A | 6 | A | 4 | A | 2 |
|  | C | 2 | C | 4 | F | 2 |
|  | E | 2 | H | 1 | H | 5 |
|  | G | 1 | N | 1 | I | 1 |
|  | J | 6 | O | 1 |  |  |
|  | M | 1 |  |  |  |  |
| Total |  | 18 |  | 11 |  | 10 |
| Industry | Ind 1 | | Ind 2 | | Ind 3 | |
| Specialization (%) | 65.2 | | 75.0 | | 80.0 | |
| Coverage (%) | 83.3 | | 54.5 | | 80.0 | |

In addition to specialization and coverage, *concentration ratios* are used to ascertain the degree of competitiveness within an industry or a market. The combined industry or market shares of a few firms are computed to show the extent of their dominance. The lower the concentration rate, the more competitive an industry or a market is. A concentration ratio of one would indicate a monopoly.

## A Particular Set of Matrices

There's no more complete application of what we discussed above than a set of two-dimensional matrices that benchmark an economy, revealing snapshots of its entire structure. This particular conceptual piece is credited to Wassily Leontief, and eventually won him the 1973 Nobel Prize in economics. Input–output tables appeared in their modern form in the 1930s but they didn't spread widely until the 1960s.[‡] This was so because of their significant data requirements for those times, as well as some ideological controversy. In the 1950s, some in the US associated this exercise with Soviet central planning and banned it! The communist regime in China rejected it too – as too western and capitalistic, though![4]

---

[‡] Earlier work was done by Quesnay in the Tableau Économique and Walras. Leontief produced early tables in the late 1930s, then he was contracted by the BLS to produce the first tables for the US in 1943. The United Kingdom started work in this area too, and the UN produced a manual in the 1950s. Canada produced input–output tables in the 1960s, and they go back to 1961.

**TABLE 5.4**

The make matrix

| Make Matrix | | Com 1 | Com 2 | Com 3 | ... | Com m | Total |
|---|---|---|---|---|---|---|---|
| | | | **Commodities ($m)** | | | | |
| Industries | Ind 1 | 0 | 20 | 0 | ... | 10 | 30 |
| | Ind 2 | 11 | 0 | 7 | ... | 4 | 22 |
| | Ind 3 | 3 | 0 | 18 | ... | 20 | 41 |
| | ... | ... | ... | ... | ... | ... | ... |
| | Ind *n* | 1 | 10 | 0 | ... | 253 | 264 |
| | **Total** | **15** | **30** | **25** | **...** | **287** | **357** |

**TABLE 5.5**

The use matrix

| Use Matrix | | | | | | | | | | |
|---|---|---|---|---|---|---|---|---|---|---|
| **Commodities** | Ind 1 | Ind 2 | Ind 3 | ... | Ind *n* | Total int | C | I | X | Total |
| | | | **Industries** | | | | **Final Demand** | | | |
| Com 1 | 2 | 0 | 0 | ... | 5 | 7 | 5 | 3 | 0 | 15 |
| Com 2 | 0 | 0 | 1 | ... | 5 | 6 | 12 | 2 | 10 | 30 |
| Com 3 | 0 | 3 | 4 | ... | 3 | 10 | 9 | 4 | 2 | 25 |
| ... | ... | ... | ... | ... | ... | ... | ... | ... | ... | ... |
| Com *m* | 10 | 6 | 11 | ... | 97 | 124 | 99 | 51 | 13 | 287 |
| **Total Intermediate** | 12 | 9 | 16 | ... | 110 | 147 | - | - | - | - |
| **Value added** | 18 | 13 | 25 | ... | 154 | 210 | **125** | **60** | **25** | **357** |
| **Total** | **30** | **22** | **41** | **...** | **264** | **357** | | | | |

The two dimensions of the tables are industries and commodities. Typically, the supply or *make table* shows $n$ industries going down and $m$ commodities going across. The total output of each industry (last column), the output associated with each commodity (last row), and the total economic output are shown in Table 5.4. Industry 1 produces $20m worth of commodity 2 (its primary commodity), and $10m of other commodities (secondary and tertiary) for a total of $30m. It doesn't produce commodity 1 but industry 2 does, as its primary commodity ($11m). Commodity 1 is also produced by other industries (3 and $n$), for a total of $15m.

The demand or *use table* reverses the rows and columns and is like a recipe, showing commodities as inputs used by each industry to produce its outputs. This is the core part of Table 5.5, in which the right-hand side column shows total intermediate inputs produced and the bottom row all intermediate inputs used by each industry. For instance, $7m of commodity 1 were used as intermediate inputs, $2m by industry 1 and $5m by all others. Industry 2 used $3m of commodity 3 and $6m of other commodities as inputs, for a total of $9m.

The use matrix can be extended with more rows, notably the value added by each industry before arriving at the total. The ever-important *final demand* can also be an extension on the right side, capturing the 'final' fate of commodities. That is, other than intermediate inputs, all commodities produced must appear as final consumption, investment, or exports.[§] Flour bought by households represents final consumption, if bought by bakers it'd be an intermediate input to making bread. The relationship between the two types of tables becomes apparent – compare the shaded parts.

Moreover, they're produced at different levels of aggregation, ranging from tens of industries and commodities to many hundreds. Even the smaller one doesn't fit on a page, and they all have many zeros. Some countries also opt for *symmetric input–output tables*, considered useful for policy simulations. Square matrices can be constructed at the product space ($mxm$) or the industry level ($nxn$). Any such exercise contains a lot of information that enables detailed economic analysis by revealing relationships among variables, such as the inputs required per unit of output. Some particular coefficients are the well-known *multipliers* used to explore direct and indirect effects on output from policies, events like natural disasters, and any other change in production processes or final demand.[**] They're also used to entertain *what if* scenarios. From output, it's a short distance to employment. Many forecasts you'll come across are based on multipliers originating in these particular matrices.

## Industry Classifications

Even from our simple examples, it's easily visible that an industry classification is really a grid. Parallel horizontal and vertical lines can be drawn at will to divide a space in progressively finer slots. They'll house an entire population. As when grids are superimposed on physical maps, we need to cut corners to squeeze some businesses into some slots. Alternatively, you can mentally visualize twisting some lines into pretzel-like shapes or akin to the squiggly international date line. Parts of a big corporate conglomerate may fit reasonably well in a slot, other parts will be awkward and require special accommodations. Either way, every unit is placed in a slot for sorting in appropriate piles. As we already know, both for granularity and aggregation purposes, each business will be put in the lowest slot possible. The choices made will affect the end data outputs.

---

[§] The simplified example here doesn't show imports, governments, and many other detailed items, for example, factor incomes, taxes, and subsidies.

[**] Revenue multipliers show interdependencies among industries and are much higher than GDP multipliers, which reflect the impact of higher output within an industry and not just the ripple effects on other industries.

Like every classification, an industry classification has a hierarchical structure and mutually exclusive categories. Primary, secondary, and tertiary economic activities are subdivided into several sectors. From agriculture and mining, to manufacturing, transportation, wholesale and retail trade, financial, business, and personal services, to health, education, and government services, and everything in between. More detailed categories are nested progressively into lower-level groupings, with businesses looking increasingly more alike. Nothing is left out. Pragmatically, such a structure is arrived at through both a top-down and a bottom-up approaches. Sometimes, as economic activities evolve, boundaries get blurred. For instance, it's become tough for some time to tell some manufacturers apart from wholesalers. Other times, strange situations may surface. Talking about allocation based on principal activity, there may be small hotels that would have long been out of business had it not been for the booze revenue generated in taverns inside them.

Industrial classifications intend to represent the structure of the economy and keep abreast of changes. Some big businesses today didn't exist even 20 years ago, others have transformed or disappeared. With rapidly changing technologies, globalization, offshoring, and outsourcing, manufacturing in many countries today bears little resemblance to that of half a century ago. New industries that reflect new products can be created based on several criteria, including critical mass and industry homogeneity. Still, most times new technologies will remain at the product space, produced by existing industries. Examples include multimedia products that came and went, for example, VHS tapes and DVDs. Either way, all classifications very sensibly include an *other* category to slot any misfit. When such a category starts to contain enough data, it becomes a good indicator that adjustments are needed.

The conceptualization and creation of industry classifications can be guided by supply or demand considerations,[5] albeit practically none is applied in a stringent sense. The North American Industry Classification System (NAICS) was the first to be based on production-oriented principles, that is, the similarities among production functions. Manufacturing sugar from sugar cane is not the same as making sugar from beets. A demand approach results in different groupings, based on the substitutability of products you find on the supermarket shelf. If consumers can't tell the sugars apart, why bother? However, homogeneity among products differs greatly. When it comes to filling the car tank, most people won't sweat it choosing any gas station – unless they collect points. Someone in the market for a door, though, can visit the workshop on the way out of town, the Windows & Doors store in the neighbourhood for more variety, the big box store nearby, or ask a carpenter to make one. Each of these door makers or suppliers will be in a different industry. Business directories or the old yellow pages will offer several business names but no industry code.

As of 1997, the NAICS replaced national classifications in Canada, the US, and Mexico. It's a classification system for establishments, and not large and complex enterprises with integrated vertical and horizontal activities.[6] The 2017 version in Canada consists of 20 sectors (2-digit), 102 subsectors (3-digit), 324 industry groups (4-digit), 710 industries (5-digit), and 928 that designate national industries (6-digit).[7] In most sectors, NAICS provides for comparability at the industry level (5-digit), although differences in the economies of the three countries prevent full comparability. Europeans uses the NACE (Nomenclature statistique des activités économiques dans la Communauté européenne). Most countries have their own, and there's an international one. Concordances have been developed to compare across countries and over time, with data from older or revised classifications. Businesses are assigned an industry code in national registers, complemented with sequential numbers for anonymization. The industry designation of businesses can change over time. (In some countries, additional unique numbers are assigned to identify businesses in their dealings with multiple authorities.)

## Additional Cast

Products add a crucial dimension to business statistics. A product is really the defining unit in the economists' conception of a *market*. So, product classifications exist, also as hierarchical grid-like structures. Because the universe of goods and services is so vast and fast-changing, classifications are balancing acts between product detail and operational feasibility. The North American Product Classification System, a companion to NAICS, contains close to 3,000 categories of goods and services at its most detailed level (7-digit). Historically, entries for goods have been more detailed than entries for services. Commonly used criteria to identify products are physical characteristics, stage of processing, technology, function, and intended use. The industry of origin also guides decisions on how to organize detailed categories into higher-level aggregates.

A widely used *commodity classification* in international trade statistics is the Harmonized System. It comprises approximately 5,300 commodity descriptions, classified in a 6-digit system determined by a variety of factors, including form and function but also component materials. Fresh potatoes are differentiated from frozen. Chairs made of metal differ from chairs made of wood or plastic. Bars of soap and liquid soap are different. Commodities come and go, the lives of many tend to be quite short. Perhaps in the future a blockchain technology will allow us to track backwards glorious old gadgets and our favourite childhood toys.

A different prominent classification is the *National Occupational Classification*. It provides another view of the make-up of employment by focusing on the occupational specialization of employees. It classifies jobs through the use of two criteria: *skill types* determine broad occupational groups and *skill levels* lead to more detailed categories. Naturally, employment based on occupations cuts across industries. The number of data scientists, for instance, can only be arrived at by identifying and counting those individuals across all industries, from IT to the government.

Except for the production of business statistics by industry, location, firm size, and other characteristics, there is significant interest in many business *activities*. These can have profound implications on the economy. Investment is an activity undertaken by most business at one time or another. Innovation is another, as are R&D, e-commerce, and many more. Measuring activities is another area where squiggly lines need to be drawn, cutting through industries and other dimensions. Depending on their prevalence or newness, some activities are subject to ongoing measurements, others may be applicable to a rare population and hard to track. While cross-economy survey designs are used, it may not be prudent to use them for extensive fishing in waters with scarce catch. The full extent of the contextual information around an activity must come to bear to design imaginative approaches. It may be easier to measure how many employees were born on February 29. With the right identification keys, employees can be found from the businesses and then linked to personal data.

## Flexible Classifications and Alternative Views

Business data based on industrial classifications have become all too common. Useful as they may be, their ubiquitous and *de facto* presence may have conditioned a boxed mindset at the expense of critical thinking related to their limitations and drawbacks. In essence, excessive focus on the industry dimension surrenders research flexibility. This is so not when data are presented by industry, but when data are *produced* by industry in such a way that other views are difficult or impossible to construct. When industry aggregates are produced top-down through sample surveys, the microdata of individual businesses are either insufficient or too scattered across surveys to support alternative views built from the ground-up.

Being one grid, an industry classification offers a particular view. Many multi-product large businesses, with complicated corporate structures, have to be squeezed in the slot of an industry code. We also saw that products and industries change, adding heterogeneity to each slot over time. Repetition of the same view may breed complacency rather than the exploration of alternative views. Researchers could contemplate flexible classifications with different allocations of businesses and craft data sets accordingly. For that, sufficient business microdata becomes a prerequisite. Ideally, internal breakdowns through the profiling of large entities should exist too.

In addition to business demographics and other uses, a good register can also facilitate spatial statistics. Establishments and their locations can be geocoded and mapped with precision. Complex enterprises can be shown on a map too, perhaps using their head office addresses – with all its caveats.

A visual geographic dimension of industries will be revealing beyond their economic size and other characteristics. Fisheries will be mostly clustered near the coasts, mining in the mountains, hotels and restaurants will be heavily concentrated in city centres, big-box retail in the suburbs. Then, next to the lists of people and businesses, consider the existence of another register, this time containing all buildings and engineering structures – some countries have it. Appropriate line drawing in that grid will separate residential from commercial, institutional, industrial, and other types of buildings. It'll identify hybrids too, where apartments coexist with offices and a coffee shop at street level. Finally, imagine superimposing this grid on top of the grid with geolocated businesses. This is exactly what a linkage exercise would do. The spatial view of an industry will come more alive. It'll show some businesses occupying one building, others many, including factories, warehouses, and retail outlets. Some others will be found in a small office, a fraction of a high-rise.

The common prerequisite among all the above is *microdata*. Not only they enable alternative views of our constructed reality, but they can combine different views leading to deeper understanding. Microdata for a population represent freedom from being hostage to a particular grid. So long as they exist, any grid imaginable can be thrown on top of them – it'll surely be populated with numbers. Moreover, microdata ease the pressure of excessive line-drawing upfront by mitigating the negative consequences of prior choices. An unspoken secret is that many of the lines we've come to know well were necessary because of our inability to deal with microdata. For a long time, microdata were shunned. Relying on slices of their distributions, like age groups or size bands, was the next best thing but inferior to the complete data sets. Microdata and their analytical potential will be critical in the future of data. They will be used to explore areas of close proximity but on opposite sides of the drawn lines. They may even redraw some well-known lines too. The more detailed our microdata, the better. Viable drawing of physical lines is best done after close inspection of the grounds. History is replete with negative experiences when borders on maps are drawn from far away. That's why I emphasized that *where* lines are drawn is worthy of debate and can be negotiable. We'll discuss microdata and related issues in more detail in Chapter 8.

## One Data

Taking a census bears little resemblance to constructing a business register. The objects of measurement are different, and so are the processes involved. A researcher specializing in retail trade would be familiar with the players in the industry, the margins fetched by different products, perhaps even the private lives of some company executives, and he'll definitely know the minutiae of placing snacks near cash registers. Notwithstanding his considerable reputation in retail, he wouldn't be called as an expert witness to sort out a dispute over how best to redesign tick boxes for binary gender on a questionnaire to accommodate LGBTQ+2S. Subject matter matters and will always matter.

In the data world, thematic areas were brought about by the incremental needs for metrics. Successive waves of new needs, and their proponents, deserved full attention rather than being relegated at the end of someone's desk waiting for an opportunity to arise. Historically, programs were set up for the census, agriculture, prices, cross-border trade, employment, output, all the way to household spending, time use, R&D, and energy consumption. Looking back, the meticulous drawing of lines divvied up everything in bite-sizes. Nowhere are the different *data disciplines* more visible than inside statistical outfits. Due to the asynchronous needs, the earlier limited technological capabilities for cross-fertilization through data manipulation, and some classic organizational inertia, subject matter silos allowed legitimate differences among data areas to get the best of their commonalities. Each program responded to the early questions posed but needs never stand still, they grow – in breadth and in depth. The layers of our conceptual constructs kept expanding by superimposing grids on top of each other, aided by the existence of a statistical critical mass by then. Core business statistics by industry and firm size from one source could be enhanced by information on their corporate structure and the characteristics of their employees from other sources. The interest in families could be complemented with information on the employers of the parents and the schools of the kids. New dimensions started to overlap, the back-and-forth had started. Implications internal to each subject matter, but also horizontal, had to be dealt with. The process of reaching out began, from the next-door neighbours to farther away. Later on, but closer to our times, the explicit realization of the feasibility of linking across multiple data sources sank in.

Given enough sources, we can navigate the data world from anywhere. *Any* starting point can trigger a chain reaction to an entire web of linkages. Census and survey data can be matched with tombstone register information, which can link to corporate and individual tax files, payroll, customs or immigration data, and everything in between. All that's needed are common keys – business numbers and personal identifiers. Workers can be connected to their employers and satisfy the fixation of a researcher or policy wonk on the welders in the auto parts industry. We can find out who they are, where they live, and where they play. While at it, we can look inside their households. We can use linkages to get in and out of business, homes, schools, and airports. There's no end. If we can start with welders in the auto parts industry and explore all statistical infrastructure, we can start from anywhere. Time to throw some distinctions out of the window. Questioning whether we expand social data by bringing in economic data or economic data by incorporating social data is pointless, really. With an abundance of microdata, what matters are good relationships and connecting bridges among all data sources.

Data *is* plural! Yet, their human connection and our insatiable curiosity make them one. Never mind how many borders the world has and how many real or imaginary lines we've drawn, we're told by those who've ventured higher up that the earth is just *one* blue planet. We've seen the video too. There's no beginning to start a trek and explore the globe, any point will do. The ostensible fragmentation in the world of data is reminiscent of physical borders – the higher up we go, the less visible they become until they disappear completely. Data are all about people and our lives. The same people who work and play, have fun or suffer, are the connecting threads. In the final analysis all data get mingled and any beginning or end is just arbitrary for their exploration. We can start anywhere, and finish where we started.

---

## Fact-Checking Tips

*Innumeracy vs. unethical behaviour*: When total employment is decomposed by industry and occupation, it offers a richer view of the labour market than each of them can offer alone. Some occupations are heavily concentrated in some industries and absent in others. This may be particularly so for industries named after the services produced by those occupations. For instance, barbers won't be employed by law firms or banks. Lawyers and accountants, though, can be employed by businesses anywhere. Not too many dentists will be employed outside of *Offices of Dentists (NAICS 62121)*, but not only dentists are employed in that industry. There are assistants, dental hygienists, and receptionists. Some occupations are widespread among many industries, such as IT or clerical staff.

Depending on the level of industry and occupational classification detail used to construct such tables, they can be both extremely informative and unwieldy. However, there's no reason for this to obscure the interplay of employment between industry and occupation. Our key message here can be delivered equally well with a 2 × 2 matrix only. Table 5.6 contains one industry (X) and one occupation (Z), with all other industries and occupations shown as composites (*n* and *m*, respectively). The left part shows the distribution of total employment by industry (columns) and occupation (rows). The middle part shows the relative proportions of occupations in each industry, and the right part shows the distribution of each occupation across all industries.

The notion of industry employment has meaning. The number 5,704 for industry X is the sum total of employees in all individual business classified

**TABLE 5.6**

Employment by industry and occupation

| | | Industries | | | Industry by Occupation (%) | | | Occupations Across Industries (%) | | |
|---|---|---|---|---|---|---|---|---|---|---|
| | | X | n | All | X | n | All | X | n | All |
| Occupations | Z | 3,245 | 1,234 | **4,479** | 56.9 | 1.6 | 5.4 | 72.4 | 27.6 | 100.0 |
| | m | 2,459 | 75,646 | 78,105 | 43.1 | 98.4 | 94.6 | 3.1 | 96.9 | 100.0 |
| | All | **5,704** | 76,880 | 82,584 | 100.0 | 100.0 | 100.0 | **6.9** | 93.1 | 100.0 |

there and accounts for 6.9% of total employment. In the event that one such business goes bankrupt, the loss of employment will be felt immediately. The notion of employment by occupation also has meaning. The number 4,479 signifies the extent of profession Z, with its particular specialization and skill set in a country. It indicates capacity and is useful to identify skill shortages. Employees in that occupation account for 5.4% of all employment. In this example, occupation Z is dominant in industry X, accounting for 56.9% of its employment. Moreover, there are only a few jobs for this occupation in all other industries, collectively accounting for 1.6% of their employment. The bulk of this profession (72.4%) is employed in industry X. (That employment in industry X, at 5,704, is higher than the 4,479 employment in occupation Z is inconsequential. Either number could be higher.)

The sum total of employment in industry X and occupation Z is bereft of any meaning. It occupies no space in the conceptual real estate, it stands nowhere, and has no denominator. Yet, you'll come across examples of such aggregations, over-inflating their 'meaningless importance'. (I've actually witnessed them.) In such a case, you need to judge whether you're confronted by an extreme case of innumeracy or unethical behaviour. This is the extent of your choices. You can picture optometrists and car dealers in the same cocktail party, thrown by the association of receptionists, but you can't forcefully comingle the business employment of those industries and all the receptionists in the country in double-counted statistics. Even if the whole thing is not meant for hefty policy but purely for advocacy, it'll backfire big time. No cause will benefit from a real-devil's advocate.

*Discontinuities vs. errors:* Classifications are meant to 'organize' our data. They've certainly become dominant in the presentation of the data and influential in their analysis. Being aware, if not knowledgeable, of how they're applied and used is material to fact-checking. By definition, classifications create groupings for a number of units of a population. One way or the other, their hierarchical nature entails increasingly higher levels of aggregation, the movement of which cannot be understood without drilling down to the microdata where the ultimate answers reside. While both the microdata and the composition behind the aggregates will be unknown, you should be aware of major changes. This is reminiscent of sensitivity tests and will help distinguish between an error and explainable causes of discontinuity.

An early question relates to the composition of classification groupings, past their constructed logic. That is, beyond the acceptance that a grouping has meaning even though the units that make it up are not the same over time. For instance, we find usefulness in the continuous existence of a manufacturing industry even though we know well that the businesses it contains today are drastically different from the 1960s. Not only there are births and deaths but, as businesses evolve, new product lines replace older ones requiring *reclassification* at some point. If this is missed, businesses may be subject to *misclassification*. In principle, taking care of such issues corrects the data. The visible changes cannot be considered errors. Such issues matter more in short-term series. A sizeable monthly decline in clothing sales may be the immediate effect of the abrupt exit of a major retailer, before others pick up the lost sales. Same for a historic spike in the value of building permits this quarter, if it included a once-in-a-lifetime project like an airport. Such compositional shifts will be known to the data producers, should be inferred by subject matter insiders, but can't be guessed by most users. Ideally, they should all be covered by adequate metadata. While protecting confidentiality, there are ways for the data source to convey the retailer's bankruptcy and the building of an airport. If such documentation is unbeknownst, both data comparisons and fact-checking become challenging and require additional steps.

Even more explicit are data discontinuities introduced by classification *revisions* and *concordances*, unavoidable for data comparisons. Following revisions, many groupings remain intact but some don't. Depending on the subject matter and the pace of change, new ones may be created, others may be combined or split further. Any change in the units of a slot will obviously affect any and all higher aggregates. Worse still are comparisons over lengthy periods. Industry data prior to the late 1990s were based on the national predecessors of NAICS. They didn't have slots for most transformative technologies of our days – there was no internet or 3D manufacturing when they were created, or last revised. Making do with temporal discontinuities is not a question of data accuracy or taking data with 'a grain of salt'. Comparisons assume a different meaning, and metadata are crucial. If they're not forthcoming, communicating with the data producer would help. In a way, moving back and forth between classifications is like translating something from English to another language or two before it's translated back to English. (Lately, though, this is getting good enough for *the gist* of the argument.)

*Maverick data:* The boundaries we draw to separate two sides and the slots we use to organize our data tend to be applied consistently. Some lines though are harder to pin down, they resist common use. The size of businesses is a case in point. Policy interest in SMEs has been a constant in every country, and governments claim to bend over backwards to support them. They've been called the backbone of the economy and engines of growth. They're counted upon for job creation. A casual observer would be excused to think that by now we'd have nailed down what exactly is an SME. However, grouping businesses by size has evaded our canny ability to negotiate common lines. Perhaps, different thresholds serve better the hosts of national programs aimed at SMEs. A small business for one may not be small enough for another.

Size bands for businesses rely mostly on revenues (turnover) or employment. Occasionally, assets may be used for some specific need. Value-added would have been intuitive too, but intensive on calculations. Revenue categories are typically delineated by some round numbers, in the equivalent of millions of dollars in annual revenues, for example, <$50m, $50m–$100m, >$100m. They also tend to exclude very small businesses, through a revenue threshold. Employment categories frequently used are 0–99 for small businesses, 100–499 for medium, and >500 for large. At times *small* starts at ten paid employees, excluding smaller microenterprises. Sometimes, *medium* stops at 250 employees. Other common employment subgroups are <20 and <50. Interestingly, such size bands remain constant over decades now. Their round upper and lower boundaries are not affected by inflation or total employment. With any definition, very large businesses represent a tiny fraction of all businesses even in large economies. In Canada, for instance, businesses with >500 employees number a few thousand and account for 0.2% of establishments.[8]

You'll come across different distributions of businesses depending on the size bands used. Moreover, such data may also be based on different business units. Distributions may refer to locations, establishments, or even enterprises – even though NAICS wasn't meant for that. Complicating matters with good intentions, an argument can be made that any sizes across the whole economy are inferior to industry-based classes. What's small in banking may be big in personal services. You'll also see such sizes too. When statistics are presented both by industry and size, the grid becomes denser. It's not always straightforward how to reconcile such statistics.

It's not the fact-checker's job to sort out this issue, nor to take sides. There are, however, steps to cut through. The lowest hanging fruit is making sure that any statistics for small, medium, somewhat large, or very large businesses are accompanied by precise boundaries and the unit of businesses. Faced with contradictions, look for points of reference to put comparisons on an equal footing. For instance, can you arrive at an estimate for firms with <100 employees from more detailed breakdowns elsewhere, say increments of 20 from 0 to 100? Then, verify the continuity of the size bands – it'd be suspicious, and certainly disingenuous, for the same source to change them from one release to the next. If it happens, complete explanations would be necessary, perhaps an infrequent redesign or an attempt to dovetail with some other initiative. An additional test, whenever feasible, could be the sensitivity of the stated conclusions to small changes in size bands. Sometimes, this can be facilitated by extra breakdowns provided.

___

## Notes

1 US Census Bureau. "Standard Hierarchy of Census Geographic Entities," *US Census Bureau*, November, 2020, www2.census.gov/geo/pdfs/reference/geodiagram.pdf

2 OECD, "Delineating Functional Areas in All Territories," *OECD iLibrary*, 2020, www.oecd-ilibrary.org/sites/ba8c7dc7-en/index.html?itemId=/content/compon ent/ba8c7dc7-en

3 GeoRef is a worldwide positioning system, based on the globe's imaginary division into 12 bands of latitude and 24 zones of longitude. It defines 288 quadrangles, equal areas with 15° longitude and latitude identified by two alphabetic characters (easting and northing). These are subdivided into one-degree quadrangles, identified by two more letters. In turn, they are subdivided into 60 zones of 1-min latitude and longitude, which can be subdivided even further in tenths or hundredths of a degree indicated by more letters. Altitudes can be added, indicated by numbers. See also, National Geospatial Intelligence Agency "Universal Grids and Grid Reference Systems," Version 2.0.0, Standardization Document, 2014, https://earth-info.nga.mil

4 Karen J. Horowitz and Mark A. Planting, "Concepts and Methods of the U.S. Input-Output Accounts," *Bureau of Economic Analysis (BEA)*, September 2006, updated April 2009, www.bea.gov/sites/default/files/methodologies/IOmanual _092906.pdf

5 US Census Bureau, "History of NAICS – Early Development Documents," Issue Papers Relating to Industry Classification, Issues Paper No. 1, *North American Industry Classification System*, www.census.gov/naics/history/docs/issue_pape r_1.pdf

6 US Census Bureau, "Introduction to NAICS," *North American Industry Classification System*, www.census.gov/naics/?008967

7 Statistics Canada, "North American Industry Classification System (NAICS), Canada 2017 Version 3.0" (Catalogue no. 12-501-X), Release Date: September 14, 2018, www.statcan.gc.ca/en/subjects/standard/naics/2017/v3/index

8 For instance, in 2019, in Canada small businesses with <100 employees accounted for 97.9% of all businesses with at least one employee. They accounted for 68.8% of private labour force, while large businesses accounted for 11.5%. In addition, in 2016, small and large businesses accounted for about 42% and 45% of private sector GDP, respectively. Government of Canada, "Key Small Business Statistics 2020," *Innovation, Science and Economic Development Canada*, www.ic.gc.ca/eic/ site/061.nsf/%20eng/h_03126.html

# 6

## The Old Guard

The world of data is being transformed in front of our eyes. From the purview of a rather specialized activity occurring in the background of economic and social life, and largely out of sight for most people who never felt quite at ease with numbers, data have become all the rage. You can hardly attend an event today or read any forward-looking think piece without being inundated with talk of data. The epithets follow suit, from the *new fuel* to the *new source of competitive advantage and wealth*. It took a massive migration to the online world, the explosive reach of social media, the arrival of the smartphone, and much commotion around *big data* to bring home the message that data are directly connected to everyday life. As this sudden conversion to the newfound truth is still sinking in, it's being reinforced by the even newer hype of Artificial Intelligence (AI). With an insatiable appetite for data, AI leapfrogged to the top of the list of forces said to make or break countries in the 21st century and decide the balance of powers.

To be fair, the ground for this data explosion was made fertile by years of efforts and many voices making the case for higher numeracy. In that sense, those involved can feel better and forget the times that felt like pulling teeth to get through to tight mindsets. There are now so many born-again converts to the data cause that, unsurprisingly, not everyone is on the same wavelength. Even what is meant by *data* is broader than it used to be, and in need of a new nomenclature – for appropriate communication. Such matters, and how they're handled, will have serious implications on the way forward and we'll discuss them. But it's not prudent to just jump into this babel-like morass, agnostic to the deep-rooted practices that carried us to the particular junction we find ourselves at. This chapter will present an overview of the data players and data practices of the past, while Chapter 7 will extend the discussion to the newer protagonists and the data they brought along. We'll identify transitions underway and pave the way for a discussion of possible future paths.

### The Players

The production of statistics largely mirrored the geopolitical developments of the 20th century. The infrastructure necessary was rolled out gradually,

DOI: 10.1201/9781003330806-6

with incremental expansions of capacity. The cast of players was finite, and their roles and responsibilities were understood reasonably well. Over time, ties and coordinating mechanisms were established among them.

## The Rise of Governments and National Statistical Offices

In the 20th century, much of the activity that led to the statistics we have come to know was the bailiwick of national statistical offices (NSOs). These institutions were set up to satisfy the growing appetite for socio-economic data by governments as they started to play a larger role in the affairs of their societies. To understand the existence and early role of NSOs, we need to understand the rise of the government sector, also a major historical phenomenon of the last century. I don't quite know all the budgetary excesses of Louis XIV's reign but in 2020 the government sector in France had climbed to 62% of the country's GDP. In the more free-market economy of the US, it still stood at 48%.[1]

Ongoing industrialization and continuous technological advancements, population increases and urbanization had no prior equivalent in human history. Economic expansion and parallel societal transformations called for more government involvement in the well-being of nations. There were heightened expectations for health, education, infrastructure, and much more. Two world wars aside, there was popular demand for safety nets and more caring societies, particularly after the experience of the Great Depression. Breakthroughs in economic thinking and the development of new tools took aim at how to go about all that. Fiscal, monetary, and social policies were counted upon to tame *animal spirits* and business cycles, and usher human societies to a new era. One in which we take control of our fate rather than be left to the whims of chance. Even libertarian types who may feel disdain for *the state* are often quick to invoke it for every problem confronting our societies. *"Where's the government?"* is a common refrain during adverse times or calamities, such as natural disasters. We, the people, ask for protection from threatening thugs and unscrupulous opportunists. We ask for the rule of law, sensible regulations, and safety inspections. We complain about potholes.

The dread of taxes for ruling monarchs and wars existed since antiquity but the ideas of universal health care, unemployment benefits, pensions, parental leave, and so much more were brand new. Someone out of sorts may still panhandle on the street, and the church or benevolent individuals can still feed and clothe those in need, but an organized society will shoulder its collective responsibilities through the state. A big, and theoretically sensible, insurance scheme. It all depends on the execution. There are many variations of this but most of the institutions we take for granted today were developed under this basic premise. So did data, because figuring out the direction, magnitude, and eventual outcomes of government interventions required quantification. Much of the official statistical apparatus was brought about in the post-WWII period.

This is particularly true for the part of the world that espoused democracy and offers a hint to the relationship between statistics and politics. Rulers of the past took the odd census and selectively engaged in some other measurements for their very own purposes, but the data collected were never meant for public consumption. For much of history, statistics for an informed public were not in fashion. *"What is not measured, does not exist"* is still applicable in many parts of the world today. From early days, it was well understood that information is power – and the quantitative component of information is no exception. In general, managing through information control is antithetical to the promulgation of data, and there is a connection between the development of statistical capacity and the degree of authoritarianism in a country. Societies with more open regimes tend to have more advanced statistical systems.

There's also a correlation between statistical capacity and socio-economic development, since the latter involves institution building. National statistical systems evolved to reflect the priorities, values, and stage of development of their countries. Some are highly centralized, like Canada's, where the NSO has a wide remit in data production – basically on everything outside of weather and sports! Others comprise several independent entities, such as in the US, where production of statistics is spread across the Bureaus of the Census, Labour Statistics, and Economic Activity, the National Center for Health Statistics, and other organizations. Like most institutions around the world, the development of NSOs has been lopsided and huge inequalities exist among countries. Some are quite advanced in their trade, with the know-how and the capabilities needed to fulfil their societal role. They're quite familiar in interacting with the research community, whether collaborating on the design of statistical programs or on the receiving end of vocal demands for more and better data. Others are housed in dilapidated buildings, severely under-resourced, and cannot possibly be expected to respond to more than the bare basics. There are countries where the mention of the statistical office in a research project unfortunately elicits puzzlement, and at times anger or laughter.

NSOs never achieved the independent status that was legally given to central banks of many countries so that they safeguard long-term monetary stability by taming the short-termism of election cycles. However, some NSOs have achieved *de facto* independence and operate at arm's length from the government of the day. There have been times when others could not withstand the pressures of their political masters and fell prey to manipulation. Recent examples came from Argentina for the CPI and Greece for the 3% budget deficit ceiling – an EU prerequisite for membership in the euro. More recently, the UK devised a new governing structure for the independence of its NSO. Even Canada, a country that in statistical circles is known to punch above its geopolitical weight, modernized its Statistics Act to give the NSO more independence in the aftermath of a couple of unfortunate incidents that led to the resignation of two chief statisticians. NSOs are still parts of their government, though, to ensure they're in tune with all happenings there and remain responsive and relevant. In essence, NSOs are technocratic institutions serving their societies at large, which justifies the term *national* (rather than *federal*, where applicable). Most governments acquiesce most of the time but it would be disingenuous not to keep in mind that not all hues of government are equally predisposed to the production and public dissemination of independent statistics.

## Additional Public Sources

NSOs represent the central nodes in the statistical systems of their countries but they're not the only data producers. Depending on the context of individual countries, and the allocation of roles and responsibilities, many other parts of governments, at all levels, produce statistics. It may be departments of agriculture for farming, health agencies for disease control, regulators for telecommunications and transportation, or the administrators of specific programs, such as the patent or the bankruptcy office. Regional and local governments also produce statistics for their areas of jurisdiction. In some cases, other entities play a role too. For instance, historically, the University of Michigan produces a consumer sentiment index in the US,[2] and the Eurobarometer conducts public opinion surveys on behalf of the European Commission for decades.[3]

## The Private and Non-Profit Sectors

Several large consultancies operate in most countries, as well as many smaller market research and polling firms. They coexist with chambers of commerce or federations of businesses and non-profit organizations, such as think-tanks and various sectoral associations. In addition to customized studies for other businesses or governments, they too produce primary data, including fee-based commercial databases, through economy-wide or sectoral surveys, focus groups, informant interviews, 'desk research', and other methods. Frequently, they venture into projections too, an area from which NSOs shy away – although there are exceptions, notably for population growth. Whether for purposes of strategy, marketing, evaluation, or financing, large businesses in particular seek proprietary information of the type that could give them a competitive edge. When some of the data produced for such studies are made public, often times they're not accompanied by sufficient methodological details. As a result, it's not possible to ascertain their quality, let alone assessing projections for 10 or 20 years out. More often than not, such data production is one-off with no fixed periodicity. There are many instances, though, that such research occurs systematically and leads to influential statistical products with lasting power.

## The Researcher 'Community'

While you'll come across this moniker time and again, many times I feel it's a bit of a misnomer. The characteristic ties that bound a community aren't present; it's more of a patchwork of networks with occasionally common interests. Individual researchers, particularly academics, have been producing a steady amount of data in support of their own research. For the most part, data are derived from experiments designed with specific purposes in mind, and they're typically of a small scale. Sometimes, though, the scale can be quite large and studies can last for years, as some medical studies track individuals over long periods. Data produced this way have fed research in every discipline imaginable for decades and have been the currency of a system driven by publications. One such study in the 1960s collected data that led to the famous *six degrees of separation*, the idea that on average any two individuals are separated by six connections at most in our 'small world'.[4] And those were the days when personal networks might have contained 100–150 acquaintances – tidying them up by 'unfriending' some wasn't yet in vogue.

There was a time that no one else could see such data, with the exception of the aggregates published. Today, journals generally require that data are made available. There are several reasons for that but much of the push relates to the contentious issue of the reproducibility of the results. Not only this is believed to be a key step in solidifying scientific advancements but there have been many cases of 'mishaps', some high profile. In numerous instances, reported results were based on faulty or inappropriately used data. Accidental errors are one thing but several cases of outright fraud have also been exposed, including making up data, discrediting those involved.

## International Organizations

There are many international and transnational organizations that compile and make available statistical data across countries. These include a number of UN bodies with *de facto* responsibilities in their areas of expertise, such as the UN Statistics Division for population and demographic statistics, the Food and Agricultural Organization for agriculture and food statistics, UNESCO for education and cultural data, UNCTAD for international trade, the International Telecommunications Union for telecommunications and information society statistics related to households. The World Bank produces and houses a variety of indicators, the International Monetary Fund is particularly active in balance of payments and international investment but also fiscal statistics, the OECD synthesizes statistics for its members states across most subject matter areas, while the statistical office of the EU, Eurostat, not only aggregates data but coordinates (the EU even *legislates*) their production across member states.

In addition, many of those organizations work closely with national statistical systems for the establishment and diffusion of standards, the cross-fertilization of best practices, and the provision of technical assistance in the Global South. In recent years, a particular push for the advancement of data has been their explicit linkage to development efforts.[5] The objectives themselves are epitomized in the form of UN targets, now known as the Sustainable Development Goals, aiming at the alleviation of poverty and other 'deprivations'. Their achievement or progress towards them requires consistent measurements and systematic monitoring over time.[6]

These have been the traditional players in the post-war statistical ecosystem. They can claim much success for the build up of the statistical infrastructure and the expansion of statistical outputs during the second part of the 20th century. In some ways, they also enjoyed a certain degree of monopoly power. By the end of the century, the landscape experienced tremors and has definitely shifted. New players have emerged in the world of data, together with new data sources and tools. The online world generates reams of data, with some assuming star status in the early part of the 21st century. Many other players and sources are surfacing too, with the potential to be even more influential. All these will be the subject of Chapter 7.

## Modus Operandi

The way humans lived in the 20th century was the stuff of science fiction in earlier times. Looking back from the end of the century to its beginning is instructive. As we rewind the tape, cars that dominate our sights and fill our streets disappear. Aeroplanes that criss-cross the skies non-stop are replaced by birds only. Widespread electrification vanishes, and whatever is plugged in drops out of the frame. On the way back then, we glance passingly at the first open heart surgery, television broadcast, transistor, and film. Then, we ended the century with the Internet, the World Wide Web, and (almost) smartphones. Successive waves of technologies cannibalized previous ones, sending many to the dustbins of history. But less heralded non-technological inventions have also had profound implications. The psychology of the unconscious mind, the Keynesian General Theory, and other breakthroughs changed our mindsets and approaches vis-à-vis individuals and society. The *survey* is right up there. It emerged as a powerful tool and claimed a first-row seat in the world of science and numbers. New disciplines, industries, and occupations were created everywhere in the world, dedicated to the business of surveys.

A good chunk of humanity's data in the second part of the 20th century was produced through surveys. They became the holy grail of research and, together with parallel advances in statistical theory, changed the way science was done and socio-economic research was conducted. Substantial infrastructures were erected to accommodate survey-taking, in conjunction with increasingly more detailed and complex methodologies to address specific issues that kept emerging. We'd be remiss not to devote a section to surveys before we discuss seriously what's in store for them and their role in the new world of data.*

## Surveys

For the longest time, human curiosity was satisfied by what could be immediately observed or 'sensed' from our surroundings. Anecdotal accounts impersonated 'data points'. Later on, to solve agricultural problems or deal with quality control and some other matters, measurements were based on multiple observations from opportunistic or convenience samples. Even the statisticians of the day were thinking that a large sample would make the estimates converge to the truth (e.g., Karl Pearson).[7] The power of random sampling had not sunk in yet. The idea of using weights to infer some truth about an entire population by surveying a tiny part was a breakthrough in waiting. Upon its arrival, rather than rely on hearsay or heed the opinions of influencers and 'key informants', we'd know for the first time what's really happening in our lives and our societies.

There were some early applications of sampling, mostly on land for agricultural purposes. Backed by some intellectual arguments, in 1903 the International Statistical Institute passed a resolution supporting the *representative method*. In the 1920s, Ronald Fisher stressed randomization, replication, and experimental design. A 1934 paper by Jerzy Neyman is credited as particularly seminal.[8] It discussed inference from samples of a finite population based on randomization (i.e., probability sampling), defined the concept of a confidence interval, and 'best' linear estimators – in the sense of being consistent and unbiased. While many of these ideas came together and were applied before WWII, the bulk of developments took place afterwards, including a whole body of methodology to support the new and growing enterprise.

---

* A disclaimer is in order here. Among the readers who know me, some will also know that I consider surveys a painful method to get to data. But good surveys are amazing tools, and I don't want to let my bias rub off to anyone. Moreover, my issue with surveys is quite particular. It's not so much the time they take and their cost, it's not begging for responses or the margins of error. My issue is entirely related to the inability of having complete microdata. (There may be a small trauma involved too. As a graduate student, I carried out a client satisfaction survey – and processed it all with a calculator. Soon after, the 'spreadsheet' appeared. You'd think I was on commission by how I promoted it.)

In the early days of the New Deal in the US, different and unreliable estimates on the number of unemployed were circulating, ranging anywhere from 3 to 15 million. A push was made for a serious measurement, and the Congress appropriated funding for an approach that would rely on voluntary national registrations. The unemployed were invited to self-register by completing a form and mailing it through the post office. Despite the massive nature of the exercise, some inquisitive minds anticipated credibility issues and, in 1937, persuaded the administration to conduct a parallel 'enumerative check census' in a sample of the areas – effectively, a random sample survey with interviews of all households in the chosen postal routes. There were some fears that sampling would have a negative impact on the reputation of the Census Bureau but, coming in the aftermath of the prediction debacle of the 1936 presidential election,[†] the approach assumed momentum and went ahead. Earlier, a survey had found that 10% of individuals accounted for 40% of incomes but the findings were criticized severely by the Chamber of Commerce that used to canvass its own members for opinions. The biggest criticism was that the sample of the survey couldn't be possibly believed. It was random!

However, a random, probability sample was precisely the killer idea. Selecting a spokesperson for many similar individuals can bring you close to the truth – which is always unknown. For this, many mathematical and statistical breakthroughs of the previous two centuries had to converge. The development of probability theory, the discovery of distributions and their statistical moments (e.g., mean and variance), the central limit theorem, the convergence to the mean (law of large numbers), as well as a variety of estimation and testing techniques, all still with us.

The sampling units in early surveys were mostly geographical areas, such as counties and census enumeration districts. The years ahead saw the development of multiple frames aiming at universal coverage.[9] They also brought about many new theoretical advances, calibrated for practical applications. From America to Europe to the 'statistical engineering' of Mahalanobis in India, such developments have filled many journals and covered a vast array of both practical and esoteric topics. From stratification and multistage

---

[†]    The 1936 US presidential election left quite a legacy. Having predicted the previous five elections correctly, the weekly magazine *Literary Digest* published a poll pronouncing Landon the winner with 57.1% of the vote. This was based on the brute force of 2.27 million returned questionnaires from the ten million it had mailed out. However, Roosevelt won in a landslide with 60.8%, carrying all but two states. The error was due to 'sampling'. Not only the subscribers of the magazine skewed more Republican but the voluntary responses came mostly from those keener on the subject. Discredited, the magazine folded within a year and a half. But George Gallup, who'd started scientific polling, correctly predicted the outcome of the election based on a sample of 50,000. His name has become synonymous with polling. However, there was more to come. In 1948, the day *after* Harry Truman won the election, the Chicago Tribune was published with the headline *"Dewey Defeats Truman"*. With many adjustments, polling lived on. Its story continues.

probabilistic sampling with or without replacement to sample rotations, panel and longitudinal surveys, analysis of variance, bootstraps, and so much more.

Conceptual and methodological advances were gradually coupled with technological ones, needed to process larger and larger amounts of data. Capturing, editing, and processing data had been challenging. Holes in punch cards represented answers to questions, and calculations were performed by hand-operated mechanical calculators. The first computer was used in the 1950 US census and by the 1960s its use had spread to surveys, gradually moving from the back end to data collection.

Surveys became more complex and expanded to more domains. Methodologies and techniques were popularized to the point that they became ubiquitous. They were conducted by statistical offices, all levels of government, private companies, non-profits, and individual researchers. They were done in high schools and even elementary schools. Today, with free software, anyone can try some version in a matter of hours.

A particularly appealing feature of surveys has been the small size required, if truly random samples are drawn. Especially when the target measures are proportions or categorical answers, choosing a confidence interval from the normal distribution, and with some notion of the underlying variability in the population, a sample as small as 400 can give estimates within ±5%. Surely, a larger sample fares always better but diminishing returns kick in very fast. With a sample of 1,000 the margin of error becomes ±3%. To go to ±1% the sample size has to increase to 10,000, and the exercise becomes much bigger. Is it worth it? In practice, as we saw, even a census could be prone to an error in the magnitude of ±2% rather than a theoretical zero.

More impressively, the size of the sample does not depend on the size of the population. The same size is good for the population of a tiny island state or China. Opinion polling typically involves smallish samples, and a tired argument among doubters is that they've never been selected. Most of the efforts are expended on the selection of a truly representative sample, the response rate, and of course the phrasing of the questions and the truthfulness of the answers. Even if all those conditions are met, election polls work until the results are very close. While people and the press have enough experience by now, a lead of 1%–2% is still reported despite a margin of error of 3%–4%. Sometimes psychology works against the brain. Except for the margin of error, we also know that the results are expected to be true 19 times out of 20. The 20th time, in the barrage of polls taken, is not exactly a black swan!

Not only did surveys produce a lot of data but they had a profound impact on governmental, private sector, and academic institutions. Their growth fed the needs of governments to monitor investments in social and physical infrastructure, and transformed social sciences and the study of human behaviours through quantification. Surveys also had noticeable cultural impacts. The lives of millions of people were touched one way or the other. There are anecdotes that early interviewers, mostly female, sought advice from reputable individuals in rural areas on where to stay, assuring the locals of their character. Loads of training and the establishment of new etiquettes were needed. More than half a century after the early attempts, I witnessed training on how to use an umbrella against threatening dogs when visiting farm houses. Countless told and untold stories exist for surveys in downtown cores, arctic areas, and jungles.

Surveys permeated all walks of life, from individuals to households, businesses, pieces of land, buildings, animals, and more. The reliance on surveys for the production of data made them influential enough that in many countries they are seen as a civic duty. Many surveys taken by statistical offices are deemed legally mandatory and, in principle, non-response by a selected individual or company breaks the law. There have indeed been visits paid by Scotland Yard to non-complying businesses. Practically, though, survey takers try to maximize response rates through persuasion and good manners.

The issue of the response burden frequently comes up in a negative light. Truth be told, 'market research' and unwanted marketing calls have become quite a nuisance today and no one wants to be on the receiving end. But there's another side to this. Even back in the early days of the growing size of the US government under the New Deal, Rensis Likert (of the scale fame) noted that one of the risks was for bureaucrats to become distant and develop an ivory tower mentality. His proposed solution? Sample surveys to keep in touch with the people. George Gallup called them *the voice of the people*. Years of experience have also shown that for every rude character with bad manners who swears at interviewers there are many more extremely polite, if lonely, individuals who can't let the interviewers go. The survey is their opportunity to talk – especially to someone from the government. In the development field, particularly, surveys give voice to the underprivileged, when no other authority asks for their opinion. It's an argument.

Surveys have also been abused. Like any instrument, there's a learning curve involved and not everyone is at the same point on it. Experience and learning-by-doing helps, and mature users have seen enough to be thrifty. But at times, excitement makes researchers see the tool more as a toy. One of my favourite examples is that of an aspiring researcher who took on a survey for the first time. He did not visualize the end output and did not use it as a guide for the design of the questionnaire, which kept growing and growing. Discreet comments (I wasn't his boss) to keep it down fell on deaf ears, unable to compete with his enthusiasm for more. Many months later, I was invited to a brief presentation of the results. The opening salvo was *"The survey collected a lot of information. Obviously(!), I cannot analyze all that. What I'll present is based on about a quarter of the questionnaire. Perhaps, I'll get to the rest later"*. That 'later' rarely ever comes.

Surveys have left a huge imprint on our societies. Despite drawbacks that we'll discuss later, they won the day. They became so common that, almost instinctively, became the answer to any data need. In the absence of superior alternatives they became the pre-eminent means of measurement not because they could measure the truth but because they could measure an interval, a range of numbers, within which the truth would lie – with some probabilities attached. But there was trouble on the horizon. Response rates eventually started to decline. Fatigue, guarded apartment buildings, telephone answering machines, caller IDs, competition for busy people's time, and a host of other impediments all contributed. Some people hide from the marketer, the pollster, or the survey boogeyman. The cell phone, especially, as it continues to substitute for fixed lines, not only affected response rates but threw off many frames, particularly of pollsters. Response rates are now plummeting. By the end of the 20th century, surveys were shadows of their old selves.

Another group of data also has had an impact on surveys. They stream from everywhere and are collectively referred to as *administrative data*. We'll explain what they are next.

## Administrative Data

Our understanding of this big component of data starts by mentally revisiting how our lives are organized. It continues with the realization that we leave traces behind everything we do. This took a boost in the online world but did not start there.

We still meet the odd elderly individuals in different parts of the world whose dates of birth are not quite known. They go by memory of what they were told. But this doesn't happen to the vast majority of people in the vast majority of places anymore. We've made sure that our paperwork is in order, from the very beginning to the bitter end. From birth certificates to death certificates and everything in between, we keep records meticulously. The vaccines given to babies after birth are recorded, the kids' registrations at school too – from kindergarten to university. All kinds of IDs, health cards, passports, or whatever else individual jurisdictions issue are all recorded. A process is waiting around the corner for whatever important we're up to. Tax filings, marriage certificates or divorce papers, drivers' licenses, car registrations, house titles, building permits, unemployment benefits, and so on. Between all the levels of government, and all the transactions we engage, such lists really have no end. Some processes are occasional or one-off, others recur periodically. All those are potential sources of administrative data.

Similar transactions take place vis-à-vis the private sector. To open a bank account, get a credit card, a telephone line or a smartphone, electricity, any type of insurance, even loyalty points from the gas station there's always 'paperwork' involved and a record left behind. The transactions executed through plastic cards offline or online, the bills we receive and pay or not, and so much more are all potential data sources. From the perspective of research, these data are qualitatively similar to those from public sources. Data are data. The difference between the two sets is in the ownership, with implications on the legal modalities related to if and how they can be accessed and used.

Our very way of life generates an enormous amount of data that have nothing to do with surveys. Potentially, they can all be tapped and used. For this to happen, several hurdles must be overcome and we'll discuss them later in this chapter. But before that, we need to go a bit more upstream. In Chapter 3, we touched upon the fact that personal identification is crucial in most transactions. All too often you need to prove that you are you. Practically, this has manifested itself in a whole array of processes in interactions with both the state and businesses. The dreaded *forms* were developed, and many people experience anguish over them. Not only has all this now migrated online but plenty more has been added. On top of standard registration information, identification requires passwords and, increasingly, proving that you're not a robot.

Almost all administrative data have their source in some *form*, whatever its name and regardless of how short or long it is. For instance, the process of getting a passport starts with an application. This particular form asks for all kinds of information, among which name, address, country and date of birth, employment and the like, even height and eye colour. Forms for a driver's license or the renewal of licence plates ask for similar information, as do the purchase of an air ticket or the booking of a hotel room. The bank may ask for our grandmothers' maiden names. At times, someone may feel 'formed out' and in a loosing battle, succumbing to such excesses. Pre-registering for childbirth to avoid a pit stop at the front desk when the time arrived, backfired. While in a real crunch, it was explained to me that pre-registration didn't mean that we can forego the registration! These are the sources where little data spring out of the ground before eventually becoming data rivers. Administrative data represent the accumulation of such individual forms.

With few exceptions, there is no coordination whatsoever among different parts of government or businesses for what is asked. Each form is designed independently, at a different time, and for a specific purpose. Governments never gained any brownie points from such practices. Imagine that after you filled in your application, paid your fee, and waited your time, one day you get your passport in the mail – together with a survey asking you if you have a passport. At any rate, the purpose of forms could be any number of things, from determining eligibility for some approval to billing, internal planning and reporting, evaluation of programs or procedures, regulatory purposes, general documentation, etc. Whatever the initial intent, it had nothing to do with the production of statistics.

But statistics can be definitely produced from such collection processes. If only for internal reporting, the passport office needs to know at the end of the year how many applications they processed, how many they approved, and how many actual passports they issued, together with all the usual *paradata* associated with their operations. They may even decide to flag which passports will expire soon and notify their holders for planning purposes (imagine)! But they don't really *produce* data. Some may publish aggregates, others may barely look at the forms further and just file them. A researcher cannot go to the passport office and ask for the number of people with valid passports, every year for the last ten years, broken down by province, gender, age, and country of birth.[‡] Neither can she go to the provincial transportation authority and ask for the number of electric and hybrid vehicles registered in each of the last five years, by make, model, and kilometres driven. The owners of these data are not in the business of data production and dissemination or serving researchers. They have not curated the data for such purposes nor are they equipped or willing to accommodate such demands. If and when such data are shared with statistical offices, and all the necessary work takes place, external research demands can be satisfied.

From the perspective of statistical offices, all external data that can potentially be acquired are seen as administrative data – even if they come from surveys other organizations have conducted! While all data are in play, some are particularly interesting. I favour a term like *census-like sources* for data sets that can be considered complete, within their own target populations. That is, the passport office has data on *all* passports, the regional transport office for *all* drivers' licenses, the local municipality for *all* marriage certificates. To arrive at equivalent totals from private sources, data from all businesses involved in an activity would be needed – with the exception of possible monopolies. For instance, *all* credit cards issued would require the data sets of all issuers, all cell phone call records would require the data sets of all service providers.

We'll discuss shortly how administrative data can be acquired and used. In the meantime, we can make some connections among the issues described so far and understand what data can be potentially available. There are parallels with surveys here. Most times a survey is identified by data that have been released publicly. However, released data represent specific choices of salient facts among endless tabulations that can be had from one single survey. Mature researchers know that, at a minimum, the possibilities extend to the whole questionnaire even though not all content may be usable (e.g., item non-response and low quality). In the case of administrative data, the form that collects the information becomes the equivalent of the questionnaire. More often than not, it contains more than what is codified in issued documents. For example, a birth certificate contains the date of birth but not

---

[‡] Researchers are free to ask anyone for data, the passport office, a bank, or an insurance company. It may be that once in a while it actually works – at some level and under certain conditions. Nothing ventured, nothing gained.

the newborn's weight. However, the weight-at-birth variable was collected in the application form filled by the parent. With the right investment, such data could become available. Asterisks for obligatory fields in today's forms also offer a hint on possible quality issues. An additional useful step is to contemplate what other data sources a given survey or an administrative form can be potentially linked to.

So, administrative sources containing specialized thematic information of the type described here could become valuable data assets. Some sources are well known but many more are quite obscure and don't easily jump to mind. Knowledge of such sources and the potential they offer can help identify alternative paths in the design of data programs, data products, or research exercises. Honing our aptitude on intelligent ways to go about data also provides insights into the dynamics of transitions underway.

## Transitions

Before the arrival of new and *big data* led to a baby boom of new converts to the cause, the world of data was already in turmoil. As has been emphasized throughout the previous chapters, statistics evolve in tandem with the evolution of the societies they serve. In other words, they adapt to the needs of their times. Examining the forces at work behind earlier transitions helps establish a perspective on what faces us today. More importantly, we'll be prepared to contemplate what may happen tomorrow – in Chapter 10.

### Speed and Higher Expectations

By the 1980s life had geared up. We got used to faster cars, supersonic jets, and loads of time-saving gadgets to the point that we started to value more speed in general. The 24-hr cable news epitomized a series of predigital developments that upped the rhythm of human movement, as well as expectations. Waiting for the nightly news or the morning paper to find out what happened was suddenly too slow. The ground was being prepared for the Internet, which soon would intensify the phenomenon of instant gratification. Not that we were ever at ease with waiting, but we had come to terms that it's a normal part of life. From the dentist's office, which has been fodder to many jokes, to menu write-ups in expensive diners explicitly spelling out the trade-off between fast food and better food, to queuing up for sports or concert tickets, we were conditioned that some patience is a virtue. We even had to wait for the weekly episode of Star Trek back then – not easy. Research involved going to libraries, searching for books, and then reading them. That took some time too. Nowadays, with everything at our fingertips and seconds away from the answers we seek, we force everything to spin faster. Same day deliveries are fully expected, even *"half-an-hour or your pizza is free"*. Notwithstanding the fact that some among us can still camp outside stores overnight for the latest iPhone, our behaviours and attitudes have placed extra burdens in many businesses, even personal relationships. *Yesterday* feels more distant today than it felt yesterday.

Matching the overall societal orientation, similar pressures for speed were put on data producers. There was a noticeable push for improved timeliness and faster turnaround of statistics. How could this be done by a system based on surveys? Surveys take time. (Some overnight polls are the exception rather than the norm.) Nonetheless, methodologies were re-examined and processes were scrutinized for time savings. A single questionnaire can be edited and beaten to a pulp but, if the outputs produced are of the aggregate type, is it worth it? Does it really impact on quality and justify delays? So, macro-editing was passed as a process innovation. Look at the aggregates and drill down only if and when necessary. Microdata were downplayed, once again. Anything to save a few days here and there thanks to popular demand. Today, in many countries, results from labour force surveys are released 2–3 weeks after the reference month. But such high-profile surveys are done each and every month, for decades. Resources are invested, teams grow experienced, and outputs get standardized. Most times it's not feasible to invest the resources and replicate that success, particularly for one-offs or surveys done for the first time. The survey instrument is still clunky. Trade-offs between quality and timeliness are present in all statistical programs, though, and subject to limits. Low-hanging fruit is all but exhausted by now, efficiencies through continuous process innovations are always worth a try but they're also prone to backfire, and cutting corners further could cross from diligence to delinquency.

This wasn't all. From *faster* it's a short distance to other Olympics-type ideals. More, farther, higher and the like. Demands for more data, *granular* in particular, started to compound the pressures on data producers, as a harbinger ushering in the new era of data. But surveys are not suitable for granularity either. Making surveys produce more data, more granular, and faster, may pass incognito on a slide in a conference presentation but will spoil anyone's day in a real office. The margin of error associated with a given sample applies to the key variables of the surveyed population. For more detailed cross-tabs, as cells run out of respondents, it grows to the point that the data are not usable. Stratification can improve the quality of domain estimates and support granularity, but at the expense of much larger sample sizes. Adding dimensions pushes the sample size very high. A household survey in the US for estimates by state and gender requires 100 strata ($50 \times 2$). Estimates from a business survey in Canada by province and territory, 2-digit industry level, and three-size bands require 780 strata ($13 \times 20 \times 3$). Such samples require tens of thousands of units. This would be antithetical to politicians asking for respond burden relief on behalf of constituents. Not to mention that costs would be prohibitive, particularly with decreasing response rates. Expanding content, increasing samples, and being faster are not exactly compatible.

Welcome to the world of administrative data! Why wait for a clunky, expensive, and slow survey that won't produce the granular detail needed anyways? The champion of 20th-century statistics started to take a pounding. Soon it would be out of vogue. From *"let's do a survey"* all heads turned to the new promise of administrative data. It was their turn to steal the spotlight and become all the rage for a while. But administrative data were not strangers to the world of official statistics. They were used selectively for a long time, mostly as inputs to statistical programs. Now, expectations were raised for using them as replacements for surveys (or parts thereof), and as the main building blocks for new programs and outputs. Over time, administrative data made more inroads into the foray of official statistics, but not by leaps and bounds. How come they didn't enjoy a bigger role in the statistics of the 20th century?

## Digitization

There are a few reasons for that, including legal and jurisdictional matters, resource and data quality issues, communication and negotiation disincentives, and some fear of dependency with no control. Progress has been made on all those fronts, and in many parts of the world. Much more is expected in the future, guaranteeing a spot for administrative data too in the future of statistics. But, by far, the most significant development came from the advancement of *digitization*. This was the *force majeure* that changed everything as we knew it.

For decades, all kinds of administrative records were stored in drawers, kept locked in cabinets, or boxed in archives. For the most part, they weren't even thought of as data. Digitization brought them back to life. But they're not zombies, since they weren't dead – just 'out'. The processes that generate them moved online and, in parallel, the old records were also digitized. This was a total game changer of the type we had never experienced. Bluntly put, with few exceptions, up to that point we couldn't handle these data. Before we're all critical, just give it a few seconds of thought – as we like to do in this book. How does a statistical office get the millions of (mostly handwritten) annual returns from the tax office so that people and businesses are not burdened with income questions? Surely, the tax office cannot ship the originals in boxes. In the very least, photocopies would be needed. This act alone would probably be enough to disqualify the idea. Alternatively, employees from the statistical office can go work side by side with tax office auditors, take copious notes, and bring them back on the bus! How well would that work? Whichever way you think of it, it was awkward. Much easier to ask a few individuals or businesses for a handful of variables that they had already submitted to the tax office. Duplication for sure. But what if it leads to less overall effort? Many 'intuitive' moves had to bide their time and wait for the arrival of digitization.

By then, the rush to substitute administrative data for surveys had intensified. Key, census-like sources were obtained and they changed the nature of many surveys. Practically, digitization is a necessary but not a sufficient condition for successful transitions. As promised earlier, we'll discuss briefly some other issues involved.

### Acquisition, Curation, and Use of Administrative Data

Collection, manipulation, compilation, analysis, and dissemination of 'official' data are governed by law. Depending on the individual country, legal frameworks may or may not cover the acquisition of administrative data. Some statistical offices enjoyed broad legal powers to acquire and even compel such data, both from the public and the private sectors, for some time. Others did not. The EU advocated for legislative changes in many member states, and several countries have recently modernized their laws expressly for the purpose of administrative data.

Moreover, no need to tell you that government has many tentacles. Different parts operate under different statutory authorities, and several issues may have to be ironed out for data to be shared. For instance, in Canada, the Statistics Act supersedes other acts, safeguarding privacy and confidentiality to the maximum. Employees sworn under the Act cannot be compelled to testify and disclose confidential information even in a court of law. Records collected elsewhere may be governed by specific legislation, such as the Transportation or the Fisheries Acts. It's quite conceivable that lawyers have to be involved at some point, even in the world of statistics. Lawyers tend to be influenced by high literacy, when numeracy is at issue. Things can get thornier quickly when it comes to the private sector. Acquiring banking transactions can invite arguments by invoking privacy concerns.

Legal matters are one thing. Practically, good manners and cultures matter more. Waving a legal text on an uncooperative bureaucrat is not a recipe for lasting success. As game theory suggests, cooperation would be ideal. For this, the right incentives are needed. Statistical offices may be able to offer useful services in exchange for the data or compensate for reasonable extra efforts needed to accommodate data transfers or sharing. At a higher level, modern governments try to promote a culture of common purpose and *whole-of-government* approaches. The complaint of citizens and businesses, why *you* ask me again for the same thing, is perfectly valid and reflects poorly on all. It has led to the mantra *"collect once and use for ever"*.

We've already mentioned that the production of data was nowhere near top-of-mind in setting up any administrative process. This brings us to the issue of quality. Survey data undergo intensive scrubbing before they're presentable. Whether correcting errors, such as missing or extra zeros, checking for consistency among responses, executing validity edits, or performing imputations for non-response, a whole series of steps are implemented. Administrative data are not subject to sampling errors but their quality poses many different issues. When forms were used, some response boxes might have been checked better than others. Some may contain correct information, others not. Who knows? They always contain duplicates, and many records have missing values. Concepts may not align. Tax forms for businesses had no use for industry codes. Names may not have been reported consistently. I have never come across one single case of administrative data not plagued by all sorts of issues. Most often, such issues surprise even their owners. Due diligence is necessary to arrive at fit-for-use data sets. This, in turn, requires a fair amount of research and two-way communication with the data owners.

In that context, good documentation and metadata become crucial. Often, however, they don't exist. In one instance, registering for unemployment benefits in the UK, homeless individuals were assigned the address of the local application office. Perhaps, the form required an address and the clerks nobly improvised to help out. But imagine the researcher who would see a spike in the number of unemployed in an affluent area without being aware of this practice? Examples like this probably have no end. There's always an awful lot more than meets the eye in such matters. Clean-up and standardization procedures also facilitate subsequent linkages of administrative data sets with other sources.

Some data may be obtained only through administrative records. For instance, it'd be awkward to survey businesses to find out which ones went under. Bankruptcy statistics exist. On the flip side, not all administrative sources may be worth our while. Quite simply, some may not be credible. Exploration and critical thinking are always friends of research. Just as an example, in South Africa not only cars but television sets must be registered. The owner must apply, get a license, and pay an annual fee. I'd like to know a lot more before I use that data set. In the US, a complex system determines the collection and remittance of sales taxes. They're up to individual states, and the legal concept of 'nexus' determines the state in which a seller operates. Online buyers in different states are not charged the sales tax. Technically, the onus is on them to report such purchases to their state tax authority and pay a *use tax*. Yeah, right!

On top of all that, there's always the unavoidable nitty-gritty of negotiations, which become particularly important if the data are to be used in an ongoing program. Issues such as the timing and frequency of deliveries must be agreed upon, together with modalities related to the format of files and technical matters. Over time, trust and a sense of common purpose need to be cultivated. The release of the quarterly GDP or the monthly CPI cannot depend on the whims of a whole bunch of indifferent individuals somewhere.

Beyond that, collaboration over the long term is needed for the future of data. In the US, several tax forms ask for NAICS codes now. But, often times, our current cultures are still such that forms are modified or disappear unilaterally, without consultation. It could be the R&D credit form or something else. Ideally, administrative programs and their feeder forms will evolve in a way that data are no longer a by-product that can be painstakingly extracted from such sources but part and parcel of their initial design.

## A Very Big Deal

Now is a good time to wrap our heads around a truly fundamental development in the world of data. It's like an evolutionary mutation that continues to gain momentum but whose genesis lies squarely in the coveted substitution of administrative for survey data. It marks the origins of a paradigm shift in how we think about data and will be a determining factor in contemplating future paths.

As digitization expanded, the rush towards administrative data intensified. With investments in efforts described above, key sources with census-like properties were initially obtained. The biggest example is tax returns of individuals and businesses. Gradually, accumulation of such sources and familiarity with the possibilities they offered not only led to substitutions for survey questions but to improvements in statistical infrastructure. For instance, business registers were solidified. Surveys were always in need of frames for representative samples. Listings came from anywhere. Dwelling enumerations, postal addresses, business directories, telephone books. Superior registers not only took over, but also found their rightful place as rich sources for actual data production. Surely, they can serve as survey frames but such auxiliary use is like using gold where something yellowish would do. This upgrade alone would have been enough justification to acquire administrative sources even if pressures to minimize survey content never existed. Even more, administrative data started to change the mindset of research designers. A long transition phase got underway.[§] Its principal characteristics are explained below.

The dominant research paradigm was based on a design that explicitly linked survey data to a set of predefined issues. Whether undertaken by public or private sector entities, or individual researchers, research always chased after some interest. There was a stated need, a set of thematic questions

---

[§] While the private sector doesn't have access to tax data or other sources subject to confidentiality, several substitutes were developed from public sources, media reports, and later the World Wide Web.

that required answers. This *research possibilities space* is depicted on the top of Figure 6.1. Using all available information, and with empirical help, vague ideas were converted to operational concepts. Ideally, the desired outputs were defined and then used as a guide for the subsequent steps of the process. The sequence of main activities included the determination of a suitable frame and the development of a questionnaire, which was tested beforehand. At the end, the answers to the survey questions would become the data that would produce the output that would answer the research questions – perhaps with a bit of extra, left for 'later'. All in all, the design revolved around a self-contained research universe with one single data source. The key feature was that the whole exercise was driven by clear and articulated research interests. Data were produced *by design* for the express purpose of feeding the desired outputs, which linked directly back to the questions. All that is shown in the left part of Figure 6.1. (Robert Groves describes this approach as having a high information-to-data ratio.)[10]

The investment to acquire, curate, and maintain an administrative data source is not driven by the desire to replace a handful of questions in a single survey. The reasons go beyond that. Surely, it can replace content in multiple surveys but it can serve other uses as well. This is particularly so for census-like sources with rich data content. For instance, business tax returns can easily replace 2–3 revenue questions in a survey but they will contain complete income statements and balance sheets. Considering the size of the effort involved, the incentive is to obtain as much data as possible at a low marginal cost. In that sense, there's a certain redundancy built into administrative data sets (lower information-to-data ratio). Some of the data may be very detailed and of the type unlikely to be asked in a survey. On occasion, a good reason may necessitate duplicate content in a survey, something that may have to pass by some corporate watchdog committee.** The main point here is quite simple. While an administrative data source may not immediately and entirely replace a survey, the longer-term accumulation of multiple sources will impact all surveys, as well as it leads to new outputs.

The collective existence of administrative sources prompted more profound and consequential changes. Extending such sources to include the new data that would soon appear – and which we'll discuss in Chapter 7 – leads to a paradigm shift. *First*, our thinking horizons become broader as we factor in the new information. What we want to study, or think we want, is no longer constrained by the limits imposed from having to collect all the data. The *research possibilities space* expands and research activities become more diversified, as shown on the right-hand side of Figure 6.1. *Second*,

---

** Another point concerns the balance between adherence to desired ideals and practical common sense. If the administrative source that can save a few questions in a national household survey is the municipality, going after tens of thousands of municipalities is not a superior option. A sincere upfront apology to the survey respondents might be a preferable course of action – until an ongoing program fed by data from municipalities is built.

**FIGURE 6.1**
Alternative research paradigms

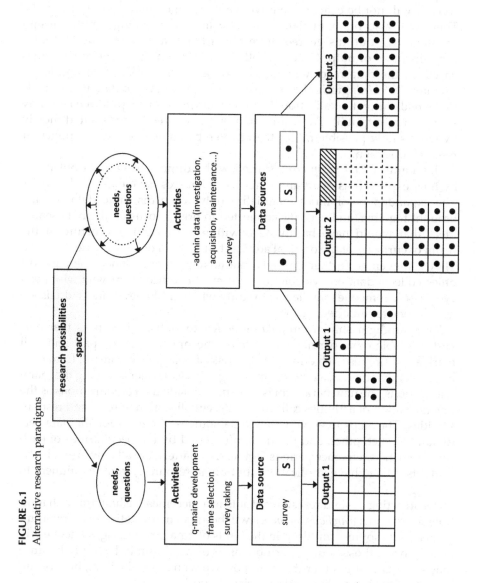

the same end output will now be produced with data from more than one data source. It will be filled in part from existing administrative data sets and from a smaller survey (output 1). (Non-survey data are represented by black dots.) The magnitude of the overall effort will depend on the relative proportions of the different parts. Many times, the contents of the new data sources will not be exactly what we want, but they may be in the vicinity. There will also be extra data. All these liberate thinking for the design of alternative outputs geared more towards the available data. Output 2 contains parts of output 1, drops other parts, but adds superior new ones based on the data readily available. (We'll describe real-life examples in Chapter 7). *Third,* gradual familiarity with non-survey data sources leads to the realization that additional and new outputs can be produced entirely from there (output 3). These are often of a new breed, precipitated mostly by what's now possible rather than driven by some overarching policy or research question.

This kind of *data prospecting* started with administrative data but it has by now exploded as many more new data sources have entered the scene. The expanded possibilities materialize only when disparate data sets are brought together, and eyeballs connected to brains focus on them. Research ideas that would have hardly been within the incentive structure or the realm of curiosity of the original administrative data owners are now at play. They can even go further through data linkages. For instance, the passport office collects data on eye colour, the transport office on drivers who need eyeglasses. Someone may decide to study whether blue-eyed individuals are more prone to imperfect vision.

This paradigm shift, from *data by design* to statistical outputs that harvest available data, permanently altered the practices of the past and will mark all future research efforts. At the risk of being too colourful, from data hunters and gatherers we're becoming *organic data* farmers. Among the many implications, the new paradigm is disruptive inside data organizations as the diversification of activities calls for a different allocation of roles and responsibilities. The steps to acquire new data sources, the need for the data sets to 'talk' to each other, and the trade-offs posed by the peculiarities of each source, such as reference periods and timing matters, are all new. As well, the skill sets needed have implications for human resources, from recruitment to training.

Constructing desired research outputs in a piecemeal fashion through multiple sources in a *warehouse approach* will become mathematically more prominent as data sources multiply. In the limits of data prospecting, someone will be looking at those sources for no reason or need in mind, simply because they exist. Looking under the lamp post we won't find the keys, but we can find something! Accidental discoveries abound. .

## Fact-Checking Tips

There are times when the veracity of numbers is not the first issue confronting us. You'll come across cases involving a disconnect between constructed concepts and their actual approximations through inappropriate or misleading measurements.

*Imposter look-alikes:* Contradictory data from different sources can be unnerving, if they're meant to quantify the same thing. Yet, such occurrences are common particularly between surveys and administrative sources. We've already discussed that estimates of the same unemployment rate concept would be different if data from different sources are used, such as labour force surveys and the unemployment rolls. Unique exclusions, omissions, and timing issues can explain away most such differences. This type of *reconciliation* is frequently a learning exercise and always helpful for analytical purposes. However, it only works when the intended target is really one and the same. It won't work if the concepts aren't fully aligned. Now, it's not unheard of that whether out of desperation or convenience we settle not for the data we want but for the data we can get – such is life. Repeated exposure and use or inertia can lead to the occasional confusion between the two.

One example comes from rates of penetration (or use) among individuals, which are conceptually the same as *per capita* measures. The denominator is really the population in both. Such concepts have a hard ceiling of 100% – since 'per capita' measures are typically expressed per 100 inhabitants. However, we've already seen such rates in excess of 100% in the case of cell phone subscriptions. This statistic is not a true penetration rate of cell phones but a crude measure that puts the activations of SIM cards in some sort of perspective against the size of a country. This choice of a numerator creates a distance between the desired and the measured concept. This is not uncommon. The key statistic of policy interest, the proportion of individuals with cell phones, is hard to come by in many countries. The availability of a supply-side number conveniently patches the hole.

With enough differences, rough comparisons can be made across some countries with this indicator. A 20% vs. a 70% would be revealing. Same for monitoring progress over time. However, this 'imposter' penetration measure breaks down when numbers are close or too high and exceeding 100%. Other than rough comparisons, only telephone companies care about SIM card sales. If policy aims at universal access, the existing indicator will never reveal the underprivileged 10%–20%. Knowing how it's estimated, it becomes clear that it drifts away from what we really want. Fact-checking the number itself is pointless, it's correct – and useless for the intended purpose.

There are many examples of the clash between desired data and measures of convenience. The owner of a bricks-and-mortar store, or a dedicated research team that he hires, can quantify quite accurately the number of individuals that visit the store in a day, track the duration of the visits, repeat visits, and other aspects of visitor behaviour inside the store. As we found out in the early days of the Internet, doing something comparable for the store's website can come conveniently from the site's log server. Plenty of data on page views, page visits, and unique visitors are captured there, and 'feel' like the store data. They're closely related but they're not the same. Page views track the number of times a page is loaded or reloaded, and many can be generated by one click. Page visits are really 'sessions' that depend on the device of access, the choice of web browser, and the duration of timed-out sessions (typically ½ hr). Unique users are not really unique individuals but unique IPs. When these are assigned dynamically by Internet service providers (ISPs), the same individual may appear under different IPs, and an untold number of individuals can be behind the same IP of a large organization.

*Shifting denominators:* Things become a little trickier when more than one number can make a legitimate claim for the denominator spot. The resulting indicators measure abstractions with different meanings. Differences can be stark or subtle. COVID-19 vaccination rates were expressed simultaneously as rates among the whole population and those eligible. They both mattered. Table 6.1 shows that the 600 vaccinated individuals represent a 60% vaccination rate among an entire population (1,000) but a more substantial 80% among the eligible population, since early vaccine approvals were meant for those over the age of 18. The rates would change again if the same number of vaccinated individuals was applicable after the vaccine was approved for the 12–18 and the 5–12 age groups. The reporting of such rates sowed some confusion as approvals for different age groups came with time lags – and they were reported separately for first and second vaccines.

Other times things can be a bit more convoluted because of partial information. Consider waking up to the following headlines, close to each other. *"Average household spending on smartwear on the rise"* and *"Smartwear prices drop"*. An analyst looking at the profitability of smartwear companies and

**TABLE 6.1**

Changing rates of vaccination

| Vaccinated (600) | Total | >18 | >12 | >5 |
|---|---|---|---|---|
| Population | 1,000 | 750 | 850 | 920 |
| Vaccination rate (%) | 60.0 | 80.0 | 70.6 | 65.2 |

the prospects of their stock prices may be somewhat puzzled. A fact-checker may salivate. An economist may be quick to explain it away based on a positive elasticity of demand. What must have happened is that the drop in prices generated a proportionately higher increase in consumption, leading to higher household spending (and therefore higher revenues for smartwear companies). Before we know anything else, this is quite plausible and may well be the case. But, it may not.

Visualize a spreadsheet that arrives at total household spending on meat, veggies, smartwear, and everything else. Subtotals for each item can be easily converted to average household spending by dividing with the total number of households. For smartwear this is shown in the shaded part of Table 6.2. With the same number of households (100), average spending increased from $384 to $504 between periods 1 and 2. This indicator is not meaningless. It can be a useful ballpark for some market research, perhaps. For more sophisticated uses, though, much is missing. The implicit assumption that all households spend on all items may be closer to the truth for some but not for others. This is always the case for newer items whose penetration is on the rise. Completing Table 6.2 with more data reveals what's really happening.

Only a certain proportion of households bought smartwear, and this increased from 40% in period 1 to 60% in period 2. With data on quantities and prices, we can see that this increased penetration alone could lead to the increase in the 'average' household spending. However, no household spent that average. Those that bought smartwear spent a lot more than that, many others didn't spend anything at all. Among *consuming households*, quantities stayed unchanged (12 units) and average spending decreased (from $960 to $840). It's not uncommon for the users not to be aware of this particular meaning of average spending or not to have the data to recreate it. Even data producers may not bother with the average among consuming households, which require their own unique denominators per item and per period – when one common denominator makes spreadsheet calculations a breeze! The economist was right for the overall market, which grew, but demand for individual households was inelastic. Their bills went down with the lower per-unit price. When you look at the 'average' consumption of beef or tofu, make your own call if denominator adjustments for vegetarians and non-tofu eaters are needed. Again, the key thing here is not the correctness of the published numbers per se, but their true meaning.

**TABLE 6.2**

Average spending by household, over time

| Spending Period | All Households (#) | Total Spending ($) | Average Spending ($) | Consuming Households (#) | Quantity (units) | Average Quantity per Household | Unit Price ($) | Household Bill ($) |
|---|---|---|---|---|---|---|---|---|
| Period 1 | 100 | 38,400 | 384 | 40 | 480 | 12 | 80 | 960 |
| Period 2 | 100 | 50,400 | 504 | 60 | 720 | 12 | 70 | 840 |

## Notes

1 OECD (2022), General government spending (indicator). doi:10.1787/a31cbf4d-en (Accessed on May 11, 2022), https://data.oecd.org/gga/general-government-spending.htm

2 The series started in 1966, see University of Michigan, "Surveys of Consumers," https://data.sca.isr.umich.edu/

3 Measurements started in 1973, European Union, "Eurobarometer: Public Opinion in the European Union," https://europa.eu/eurobarometer/screen/home

4 There are volumes of literature on this but one key article that made it famous was published by Stanley Milgram, "The Small World Problem," *Psychology Today* 2 (1967), pp. 60–67.

5 The World Bank's flagship publication "World Development Report 2021: Better Data for Better Lives" was dedicated entirely to the relationship between data and development. See Report at https://wdr2021.worldbank.org/

6 There are 17 goals, each with many detailed targets and a plethora of possible indicators. United Nations, "Sustainable Developments Goals," www.un.org/sustainabledevelopment/sustainable-development-goals/

7 Interesting stories on several of the developments discussed here are contained in David Salsburg. *The Lady Tasting Tea: How Statistics Revolutionized Science in the Twentieth Century*. Owl Books, 2002.

8 Jerzy Neyman, "On the Two Different Aspects of the Representative Method: The Method of Stratified Sampling and the Method of Purposive Selection," *Journal of the Royal Statistical Society* 97, 4 (1934), p. 5, www.jstor.org/stable/2342192. Interestingly, in the paper itself, Neyman credits Professor A. L. Bowley for many of the developments for the 'representative method,' and cites papers of his going back to 1913.

9 See, for instance, Morris H. Hansen, "Some History and Reminiscences on Survey Sampling," *Statistical Science* 2, 2 (1987), https://doi.org/10.1214/ss/1177013352

10 Robert M. Groves, "Three Eras of Survey Research," *Public Opinion Quarterly* 75, 5 (2011), pp. 861–871,Oxford University Press, https://doi.org/10.1093/poq/nfr057

# 7

## The New Era of Data

Following the footsteps of past technological breakthroughs, digitization became the enabler for widespread advances. The diffusion of computing power was fast and furious. Pieces in cumbersome legacy processes were re-arranged, changing the business models of all sectors. Just-in-time inventories were followed by integration of back-end and front-end systems in manufacturing, wholesale, and retail, tracking every single item. Radio-frequency identification linked transportation and logistics with global supply chains. Payment systems expanded to facilitate debit transactions and, later, smartphones too did the same.

The public arrival of the Internet and the World Wide Web was just around the corner. Their mainstreaming gave a big boost to a new *online* world that kick-started so much more. By the late 1990s, websites were proliferating and embryonic e-commerce had started. All that was soon intensified with Web 2.0 and the arrival of many big name players we know today, including social media. In parallel, successive bouts of innovations transformed the cell phone to a smartphone. Each technological wave added devices, functionalities, and possibilities. The Internet of Things is said to explode with 5G and beyond, something which is only beginning.

All new processes set in motion under this frenzy of activity generate data. These *digital footprints* are embedded in the innards of digital technologies. Businesses, governments, and the research community quickly realized that data unthinkable in the past are now within reach. The old data neighbourhood was marked for gentrification. By now, it's unrecognizable to visitors from the past. Many new players generate vast amounts of new data, which come in a variety of shapes and forms. Some are data that we always wanted to have. Others didn't even exist in our imagination.

### On New Sources and Methods

This section will briefly outline the contours of the new data, grouped by source for exposition purposes. Of immediate interest is their nature and qualitative features, and we'll steer clear from quantifying their volume. Suffice it to say that they're all already huge and, unlike the discrete

DOI: 10.1201/9781003330806-7

increments of survey data, the number of data points associated with any particular source increases continuously. Collectively, they're poised to grow by leaps and bounds.

### Transactional Data

Detailed data associated with fixed-line telephony existed for a very long time. They were indispensable to the industry's pricing structure, billing requirements, and the settlement of accounts among companies nationally and internationally. They also supported demand analysis internally, while the data shared with regulators or statistical offices aided the monitoring of national policies, such as universality. The fast build-up and near-complete coverage of *wireless networks* brought along large amounts of new and unique data, particularly in conjunction with the smartphone. Telephone exchanges produce *call detail records* that log *metadata* (another meaning here), such as the time and the duration (timestamps) of any transaction – whether a voice call, text message, or Internet access. Mobile antennas are housed in cell towers with unique IDs and geolocation coordinates that allow for geographic granularity. Network records contain domestic subscriptions for inbound and outbound roaming. Many eyes are on such data for some time for their research potential. In addition to connectivity questions that have been smack at the centre of much research in recent times, such as the frequency, intensity, and type of use, these data can support research in mobility – even domestic and cross-border tourism. Their use shot up during COVID-19 to shed light on population movements under imposed restrictions.

Integration of systems connects the movement of goods from stocked shelves and cash registers to warehouse deliveries in an endless loop. From big box stores to smaller places, real-time tracking is further enabled by barcodes and scanners. We're at the point of self-scanning and, who knows, perhaps soon barcodes can be read from a distance and our purchases displayed on a screen in the cart or our smartphone. Then, we can just wink or smile and walk straight out of the store, in exchange for a debit in our account. Whichever way, what goes in retail stores, what gets out, when, at what rate and price, are all recorded in excruciating detail. Point-of-sale *scanner data* can be sliced by month, week, day, and time of day. Even the method of payment is captured, complete with credit or debit information. Is it any wonder that such data, if they can be had, are seen as an alternative to sending people to supermarkets and record a handful of prices every month for the CPI? Consumer spending surveys that determine a representative basket of goods and services come into focus too. So do retail surveys. Other than having ongoing access to such data, new challenges need to be overcome. Universal Product Codes behind barcodes are extremely detailed. Different cans of tuna have a different one. Conversion to product classifications is necessary for statistical production.

Increasingly, there's no need for physical visits to homes and businesses to read utility *smart metres*. Thanks to a spin-off technology from the telephone caller ID feature, wireless meters installed by electricity, water, or gas companies can measure consumption remotely through radiofrequency electromagnetic fields that transmit data in short bursts. Smart meters communicate directly with the central system helping optimize network performance by monitoring consumption flows throughout the day. In the process, utilities become repositories of very detailed data too. These could be invaluable in research that has been practically scratchy, such as energy consumption in households, businesses, or buildings.

The growth of *e-commerce* platforms generates oodles of data. Every purchase is recorded and every click is examined, even our browsing history. Amazon, Alibaba, e-Bay, and other giants not only have become repositories of big data but they make extensive use of them. Data analysis is literally driving their business. Unlike the previous examples, their customer data are global. Somewhat similar data surface from the *gig economy*, particularly in markets with a few dominant players. An early example was Amazon's Mechanical Turk for micro-tasks. Today, well-known examples are Uber and Lyft in transportation and Airbnb in accommodation. With access to such data, researchers can create early profiles of new phenomena before official statistics catch up.

Amazon, Facebook, and Twitter were not the first companies to have and use customer data. Google wasn't the first to know that you visited the corner store or your local pub. Earlier forms of payment, though, put a lid on things. Now all transactions in the new platforms are ultimately linked to bank accounts of individuals or businesses, creating another significant layer of data. The use of 'plastic' money and other electronic means of payment have expanded to the point that many economies are approaching cash-free status. Banks and other financial institutions know many of our moves and amass an astonishing volume of *transaction records*. Even a pack of chewing gum or a cup of coffee is now recorded. Data have become more than a sidekick to their main business, and financial institutions can easily hire talent to analyse them. On the other hand, each financial institution holds only a piece of the puzzle. Also, *fintech* innovations cross and blur sectoral boundaries, likely segmenting such data further.

These are examples of notable pockets of brand new data or data reincarnated in new forms. The connecting thread among them is that they're generated through recorded transactions, which now leave a clearer trace. Unlike census-like data, each individual source does not represent any total, except perhaps in isolated monopoly cases. Totals would require aggregation across *all* businesses in a sector, for example, all wireless providers, utilities, banks, or grocery stores. But then, in many countries such industries tend to be concentrated, with a handful of firms accounting for most of the market.

One way or the other, private sector companies have always had and used internal data for their purposes. Banks have always had research teams and have been big users of all kinds of data in their modelling exercises. Telecommunication companies have had statisticians working on econometric modelling for decades. Guinness, the brewery, established a reputable statistical department more than a century ago. Most major companies, from manufacturers to pharmaceuticals, have had such research units for many decades. In fact, much progress in applied statistical theory has been generated in such places.[1] Historically, they cross-fertilized knowledge and best practices with academia and governments. There are at least two noticeable new developments, though. First, the size of internal company data has exploded and is expanding by the day. They also reside in structured databases, something conducive to ushering in data analytics. The old client satisfaction surveys can no longer compete. Neither can research on product uptake 2–3 days after an advertising campaign. The level of *beta tests* performed at lightning speeds has no precedent. Second, the mere knowledge of the existence of such data offers tantalizing possibilities and whets the appetite of outsiders. Researchers and governments would certainly like to put their hands on them. We'll explain shorty how this can lead to tensions.

## Log Server and Web Data

The architecture of the Internet is such that data can be captured at several layers. Network servers log massive amounts of data at key exchanges, including at the Internet service provider (ISP) level. Except for data related to the technical performance of the network itself, detailed two-way data traffic data of 'business value' are also generated. In the early days, classic marketing concepts clashed with the online world and created confusion. What the new technology was measuring was interesting but not what was sought. For instance, organizations with early online presence naturally required data on unique visitors, their characteristics, and patterns of use. That proved to be a difficult question for the new medium. Statistics produced by *log servers* captured visits by unique IPs rather than unique individuals. IPs were assigned dynamically to subscribers by local ISPs, meaning that the same individual could appear as several different IPs. On the flip side, an entire organization's IP was masking many unique individuals whose visits went through the same pipe. Page views became another controversial statistic as no one was sure if the 'viewer' actually paid any attention to the content. This had its parallels with TV commercials. Eventually, clicks came to dominate as the statistic of choice for ads.

Volumes of data can be harvested from the web for any number of research purposes. Different businesses make available different pieces of information, ranging from company structures to product specifications and prices, to financial and employment data, or community initiatives.

Piecemeal collection of such data requires effort. Still, the previous alternative would have been a business survey, which is no bargain either. *Web scraping* automates and facilitates such data collection, with a crawler doing the bidding of the researcher. Such methods may feel sneaky at times but businesses do choose what information they make public. If scraping is used as a faster replacement for slower manual extraction of information, there's no sinister purpose. As any approach, it has its pros and cons. Certainly, extra vigilance is needed when it comes to the credulity of the data collected. We know well by now that in addition to data nuggets the web is full of bad information too, including *deep fakes*. From businesses' data to peoples' faces! Triple-checking and verifying through any means possible is a must.

## Search Engines and Social Media

In principle, data in this space are just amassed in platforms, virtual places where people can frequent and mingle. There's no need for an overt upfront payment. Search engines don't even require registration – until expansion in services elsewhere blurs the lines. Google, for instance, is present in email, maps, plays, and cloud services. Whichever degree of 'purity' individual platforms have, with or without registrations or commercial transactions, they all have a good idea of who we are and what we do there through several means. They surely do generate loads of data about us. Now, there are volumes of literature explaining that there's an awful lot more involved here – from the vagaries of human nature, our wiring included, to manipulating algorithms.[2] Society, governments, and the platforms themselves don't quite know the best way forward. But we'll just stick to data.

Search engines provide pathways around the Internet labyrinth. The success of Google turned its name to a verb – at no time compared to Xerox. Being the search engine of choice, the data it captures and accumulates reflect the interests, concerns, and curiosities of the world. There are always arguments on the meaning, appropriate uses, and limitations of such data. For instance, other than enriching our daily vocabulary, it's not clear what was accomplished by the flood of queries for 'malarky' during the US vice presidential debate of 2012. Nonetheless, the allure of the new source made some researchers salivate from early on. Even if only back of mind, there's a lingering attraction to the sheer volume of data, which can be accessed via Application Programming Interfaces (APIs). Of course, even though its inventors created a search engine, Google realized that it's also a data company. It created its own statistical operations, crunching numbers by aggregating millions of queries. It produced research that was plausible and credible. Until it wasn't. Predicting the spread of the flu faster than the Centers for Disease Control and Prevention (CDC) backfired (see Fact-Checking Tips at the end of the chapter). While this section discusses data availability and not research methods, in fairness, erroneous predictions can be had from any data source. Google remains by far the dominant search engine today, with the exception of Baidu in China.

Registrations on Facebook are only the entrance to the data piling up there. The level of activity concentrated under one platform has no historical precedent. From text to audiovisual materials, the space is bustling with content. Quoting from Facebook's privacy policy, data include "...*the types of content you view or engage with, the features you use, the actions you take, the people or accounts you interact with, and the time, frequency and duration of your activities*". The collection extends to "...*information from and about the computers, phones, connected TVs and other web-connected devices...and we combine this information across different devices you use*".[3] Many users also volunteer additional profile information, such as racial or ethnic origin, political views, and religious or philosophical beliefs. As if all that is not enough, Facebook also gets information about users' online and offline actions and purchases from partners and 'third-parties'.

The International Telecommunications Union (ITU) reported that at the end of 2019 half the world's population was connected online.[4] Reportedly, Facebook's following wasn't that far off! The platform also alleges that the majority of users use it daily. The data thus amassed are humongous. In a couple of months, they can outsize all the data of some statistical outfit over the past century...that kind of scale. To make sense of them, and expand its business, Facebook's statisticians crunch numbers constantly. The platform even collaborates with the OECD and the World Bank on joint products, as if to flag its prestige in the data world. When its huge societal impact had first started to sink in, parts of its data holdings began to be made available through APIs to the research community. Without a playbook, how exactly to go about this is still work in progress. Earlier models, particularly for access by developers, are becoming more restrictive in the aftermath of the Cambridge Analytica debacle that compromised the personal data of millions.[*]

Twitter started differently, purporting to connect ideas rather than people. A current limit of 140 characters per tweet attempts to instil some discipline. Popular conversations can be short-lived but threads can last for a long time. Although Twitter's reach falls short of Facebook's, it's still massive by any standards. Moreover, not only it exerts significant influence on the diffusion of news but it has emerged as a news source on its own right. There's even talk of *ambient journalism*, the always-on flow of news. In a few years of operation, Twitter also sits on top of mountains of data. In addition to what's collected upfront, users can choose to share emails, phone numbers, locations, and contacts. Combined with other information these can determine age, gender, languages spoken, and other user characteristics and preferences. Tweets, retweets, likes, content read, and lists of people following each other

---

[*] This happened around the US presidential election of 2016, see Fact-Checking Tips at the end of this chapter.

are all recorded. Even with no registered account, user log data are collected, including access device, IP address, browser type, operating system, pages visited, and so on. Twitter too receives information from third parties, including advertisers.[5]

Twitter also mines extensively its volumes of data. Responding to growing demands from researchers in many disciplines, it makes some data accessible – for free or though 'premium' fee-based options. For example, APIs can return a random sample of 1% of all tweets worldwide – still in the millions per day – or filter for more specialized data through user accounts (handlers), selected words, and geographic areas. Considering the large number of links involved, some studies employ big data techniques to sift through.

These are the poster children behind the *big data* that generated much buzz in the early going of the 21st century. Besides their main business, search engines and social networks have become data powerhouses. Differences among them by area of specialization or groups targeted (e.g., Apple and Spotify for music) coexist with a certain convergence of popular features and functionalities across the spectrum. There's certainly overlap in the people participating in these platforms. The likes of Yahoo, Bing, TikTok, Reddit, Snapchat, Instagram, and YouTube are among us today but none has had a very long life yet. We don't know what the future holds but the phenomena they represent aren't going anywhere anytime soon.

## Earth Observations

In 1957, Sputnik 1 became the first man-made object in space. It only lasted a few weeks and had no camera. Now, it's quite crowded up there. Government and privately owned satellites spin around the earth in large numbers. Many are geostationary and in low orbit, others much higher. Some are large but others come in small cubes of 10 cm$^3$, forming networks. Their cameras scout the earth constantly, and at increasingly higher resolutions. Satellite pictures have already produced huge volumes of data used in multiple applications. Detailed measurements of atmospheric temperatures and humidity are used in weather forecasting, climatic and environmental measures. Surface measurements are used in vegetation and biomass analyses, track the status of greenhouses, and assess yield damage from diseases or pests. Statistical applications already include replacing entire crop surveys. They can even be used to estimate the content volumes of reservoirs by detecting thermal properties or shadows.

Any earth observation from above is conceptually in the same space. Helicopters employed by TV stations for traffic are falling from fashion but unmanned aerial vehicles are becoming increasingly popular. They're particularly adept at measuring physical infrastructure and ideal for before-and-after situations in natural disasters. Many more uses will become possible as algorithmic approaches convert images to usable data. Their impact is being felt. It already sounds weird to suggest a survey scheme for the measurement of solar panels on rooftops – but that was exactly the approach a few short years ago.

## Sensory Data

All kinds of physical devices that detect signals of interest can be placed in friendly and hostile environments. They're mostly blind but they compensate with other senses. Proximate or remote sensors are already used extensively in healthcare, agriculture, transportation, and inside our homes. Wearable medical devices are incredible data generators. A sleep apnea machine can generate and transmit enough data during the night to keep someone busy for days. Car insurance companies use devices in vehicles of consenting drivers. Autonomous driving is all 'sensory'. Many manufacturers are becoming data-also companies. From telematics and GPS to ground speed radars and liquid level sensors you can excuse an older farmer for mistaking John Deere for a tractor company.

Much like with surveys earlier, new methodologies are being developed for the processing of sensor data. Such work ranges from how to identify and clean errors to what techniques to deploy in their analysis. Rather sampling, non-response bias, or analysis of variance, statisticians must now grapple with new data matters. Unstable wireless connections, network congestion, outages or battery lives, blockages of any kind, or simply material degradation can all produce anomalies. Sensor data too have missing values and outliers, values stuck at zero, but also *drift* (values that deviate over time) and *noise* (dispersion, part of the signal). Data from sensors are expanding and they accumulate fast but, allegedly, we *"ain't seen nothing yet!"* 5G and subsequent technologies are said to propel matters to another level. In the Internet of Things, sensors with unique identifiers embedded in anything imaginable will transfer data non-stop through wireless networks. The future may well be inundated by such data, ripe for the picking.

In addition to new data sources, our overview extends a few comments to *crowdsourcing* and *open data*, which represent *methods* of generating and disseminating data, respectively, in the new era.

## Crowdsourcing

A portmanteau of crowd and outsourcing, it can encompass activities in any area where 'the crowd' is called to pitch in, such as funding projects or undertaking micro-tasks. The basic premise behind crowdsourcing data is that the power of local resources, typically volunteers, can substitute for expensive centralized collection or it can actually produce usable data where no other method can. Leveraged at the right place and time, it can prove particularly useful in conditions involving localized needs and hard-to-get data, even if data validation will surely be challenging.

## Open Data

In the spirit of other open movements, open data generated quite a following in the early years of the 21st century. In a narrow sense, the term refers to dissemination of data that can be used, shared, and re-used freely. Other than attribution, no restrictions are placed on the data. In that sense, open data became part and parcel of the newer thinking on copyright and free licenses, which are royalty-free, non-exclusive, and perpetual. A particular one is the *creative commons* licence, which allows the free distribution and derivative uses of data even in the presence of a copyright. This form of *copy-left* only requires that new creations based on the original work also grant the same liberties. Early advocacy was particularly targeted on government data since their production was financed by public funds. Governments at all levels were quick to oblige, setting up many open data portals. However, in a broader sense, the open data movement is much more than a means of dissemination and went well beyond government data. Non-profits and NGOs entered the arena. Expectations were raised that with open data 'citizens' would answer all their lingering questions, from the use of their money as taxpayers to socio-economic, ecological, and environment issues. With the arrival of the new data and the newfound interest in all data, open data became the darlings of development agencies. Their mainstreaming involved a sectoral emphasis too, creating interactions with the business world. Even euphoria was allowed to set in. Underpinned by open data, new technologies will enable the building of services to answer questions 'automatically'. In any event, proponents of open data can realistically point to many successes, much has been achieved in a short period. *"Thousands of businesses around the world owe their existence or their growth to the release of open government data, and hundreds of civil society organisations have embraced open data as a key element of the social change toolkit"*.[6]

Still, considerable ambiguities surround the boundaries of open data, not the least of which is the concern over confidentiality. NSOs can argue that all the data they historically released are open data. However, aggregate data are not that helpful to the promises of open data. Microdata are needed – that can be manipulated and re-aggregated, including in real time. Apps showing the arrival of buses based on the detailed scheduling of transit authorities don't work. Waiting at a bus stop for a specific route, two buses that are supposed to be 15 min apart may arrive back-to-back long after their expected arrivals. Only when buses are equipped with GPS, and the GPS is turned on, apps with the real whereabouts of buses become possible. Governments, notably at local levels, release increasing amounts of open data, seemingly unencumbered by statistical laws – data weren't collected under such auspices. This creates precedents, as well as raises future viability questions. Mirroring recent setbacks in other *openness* movements, the reality that gradually sets in is one of the tensions between 'free the data' and 'free me' from privacy intrusion. Inevitably, the issues of open data merge in lockstep with the issues of all data.

So, the old data neighbourhood has become an unrecognizable place. It's bustling with activity, eccentric names, new attitudes, and plenty of opportunities. Yet, in all its modernity, it poses navigational challenges with several corners suspect to an aura of transience. Repeat visitors from the past would not know their way around but to newcomers it would all look and feel as it should be. They too must know, however, that the same will happen to them. Any nook they discover today or picture they take will be relics soon. Just wait.

## Lessons from the Early Go

The new data outlined in broad strokes above are put to local use by their owners for 'business purposes'. Unsurprisingly, demands for access are also made by governments, other businesses, international organizations, and academics for 'research purposes'. Where this has been possible, books and thousands of articles have already been published. The process of learning about new sources, methods, data peculiarities, conditions of use, and general *dos and don'ts* has begun. However, the learning curve is steep. Each data environment is different; there are few commonalities even within 'social media'. Some of the new data come fragmented in tiny pieces. Such *nanodata* will need assembly even to arrive at the unit of microdata. Mention of these challenges, though, doesn't mean to discard the possibility of future data unification under one roof – someday. After all, our current predicament was something of a long shot a couple of decades ago. An open mind fits well in the mix of qualities needed to contemplate the future of data. Perhaps, audacity too.

Overwhelmingly, the new data are privately owned and contain sensitive personal information. As a rule, they can't be accessed by outsiders. Any access will invariably entail stringent protocols, and anonymization is a must. While a case can be made that a fair amount of these data has been provided voluntarily and willingly to their current owners, the fact remains that some were provided more reluctantly than others and they're all subject to reading and understanding the infamous *fine print*. Also, some data are 'harder' than others. Sentiments and opinions, likes and dislikes are more pliable than cold financial data. Before we bring old and new data together and look at their combined implications on the design of research, it'd be useful to carry with us a few comments on the new data sets that have already established some track record.

### Keep an Eye on These

Research based on data generated by search engines and social media started internally but soon expanded to many disciplines with a focus on human networks and behaviour, business realignments, even predictions of phenomena. Historically, it's still early days and experiences accumulate. Improved use will benefit from better familiarity with the new frames, sampling methods, processing techniques and, eventually, interpretation abilities. Meantime, bear in mind the following.[7]

*Representativity:* Both the volume and the newness of data from search engines and social media have mesmerized many researchers. The number of individuals behind these data is truly staggering. However, we can't allow any implicit underlying assumption that sheer size can compensate for their lack of representativity. We'd seen that before. Learning new things cannot cloud our judgement and forget what we'd learned more than a century ago. Inferences from such data for societies at large, without adjustments, are risky.

*Platform-specific biases:* Analytical inferences can only extend to the population of each platform. Therefore, studies of 'human behaviour' are really studies of the behaviour of the users of a specific platform, whose composition differs from the others. Moreover, data cannot be aggregated across platforms. We're nowhere near the point of identifying individuals across Google, Twitter, and Facebook. Studies to date use one data set or the other. Knowing the demographic make-up of platform users is crucial, and stereotypes abound. Some are more appealing to younger adults, some more to some races, some to more affluent or educated individuals, others to kids. Moreover, we don't know how they'll evolve.

*Psychosocial conundra:* People's behaviours may differ across media due to echo chambers, bandwagon and group-think effects, role playing, or other reasons. Generally, Facebook users tend to portray their lives as they would like them to be. Google searches may reflect more truthfully what people are looking for. Carl Bergstrom and Jevin West in *Calling Bullshit* describe this as *"what you see depends on where you look"* and, comparing results from the autocomplete features, state that *"It seems that people turn to Google when looking for help, and turn to Facebook when boasting about their lives"*. To compound the matter, individuals can have multiple accounts. Twitter's privacy policy states that users can create and manage multiple accounts *"to express different parts of their identity"*. If we could ever combine data across platforms and identify individuals, it seems that in addition to human behaviour we'd study split personality disorders too. Any research must be fully aware of such issues. But there's more. Many data points are even more suspect.

*Misinformation, disinformation, and lies:* "*On the Internet nobody knows you're a dog*" comes to mind. It's well documented that plenty of bots are passing as humans in social media. It's been reported that only the number of such accounts actually purged amounted to billions![8] There are also individuals with fake or constructed personas, at times professionally managed, to create a certain image or exert influence. Again, at present there are no handy corrections for such biases.

*Temporal distortions:* Additional issues creep into the data over time and introduce further biases. First, the number and composition of users follow the ups and downs of the popularity of platforms, influenced by serious events or simply rumours. Second, platforms frequently make changes or update features for various reasons. These are particularly important when it comes to personal data or widespread network practices. For instance, Facebook instituted special measures to weed out misinformation before the 2020 US presidential election but relaxed them shortly after. Such decisions lead to repeated data discontinuities. So, *"you can't step in the same river twice"* is probably applicable to data sets obtained at different times.

*Technical methods:* The algorithms used on publicly available data feeds can be open-source but also proprietary or undocumented. As a result, different researchers obtain and use different samples. For instance, filters used on APIs for Twitter may yield unstable proportions of tweets. Such matters are made worse by *embedded researchers* in a platform, who may be unable to reveal their data or corrections they made. This poses transparency and credibility issues as it clashes with journal rules for public data sets and hinders reproducibility. Considering that negative findings are rarely published, an alternative for demonstrating robustness of findings would be running a study on two or more distinct data sets (e.g., collected at different time periods, using different methods, or on different platforms). That may be a tall order but some findings will be suspect to *"running the right method at the right place and the right time"*.

*Medium-specific matters:* Text, still images, audio, and video are increasingly converted to statistical data, in part to satisfy the insatiable appetite of AI. Many mishaps have occurred in neural network research, most often attributed to faulty data in labelled corpora extracted from such media or their incorrect use. Photographs used to extract information that would help identify humans, animals, or specific objects can be mislabelled. The same can happen with audio. Text has its own unique features. Twitter is particularly well-suited for textual analysis of current interactions. Network or sentiment analyses harvest data from media and journalists' handles on certain topics or hashtags. Then, researchers either manually code the content of tweets or, given how massive they are, use automated approaches. Such automation is insensitive to context or linguistic complexity and treats text as a bag of words. The new tools are accompanied by new terminology too. Splitting text into a set of words is *tokenization*, unifying words that should be identical but are written differently is *term normalization*, while the removal of multiple occurrences is *deduplication*. Before instructed to look for meaning in chosen words, this type of processing cleans text some more by removing unwanted noise. (From Cornelia Brantner and Jürgen Pfeffer in "*Content analysis of Twitter: Big data, big studies*" I found out that 25% of English texts consist of only 17 words, like *the, a,* and *is.*)[9] Additionally, research methods are optimized for English language texts, while languages with more complex inflections and word composites need extra manual tasks. Common graphic outputs from such work include visual representations of lists, such as *word clouds* with size proportional to their frequency.

*More stuff:* The detailed list of metadata relevant to the use of the new data can be quite lengthy, and material to their interpretation. For instance, Google stores final searches not as the text was actually typed by the user but after autocompletion. Twitter dismantles retweet chains by connecting every retweet back to the original source rather than where it was triggered. Searching for tweets in certain geographic areas may be biased by the fact that only a fraction of users allow the addition of geographic information. The more such details are documented and known among researchers, the better the end outputs. An interesting corollary of the new data is also that the remit of the research they enable can be international. A researcher anywhere can access data for any country or many countries at once much easier than ever. This too has implications, as national research was usually the purview of national researchers and was fed by nationally grown data.

There's a time and there's a place for all new data. From the experience to date, there will be many times and many places for useful things to do, questions to answer, and facts to learn. There will also be times when limitations may not be limiting! For example, the lack of representativity in Google's *popular times* data will not be an issue when it comes to traffic data. On the contrary, sheer numbers are exactly what's needed to show how busy the roads or the places you plan to visit are, and at what time.

## Research (Re)Design

With exposure to old and new data sources, it's time to take research design head-on. This task is becoming trickier by the day and complicates the working lives of many. Yet, it's also becoming more exciting than ever – it's no longer a matter of finding the 'right answer' (e.g., running a survey). Balancing over a whole lot is the name of the game. We can start where we left off in the previous chapter and, to help our discussion along, a sketch of the new research paradigm is reproduced as Figure 7.1 – with minor modifications. It was explained that even before the arrival of all the new data to the scene a

**FIGURE 7.1**

The new research paradigm, supported by multiple data sources

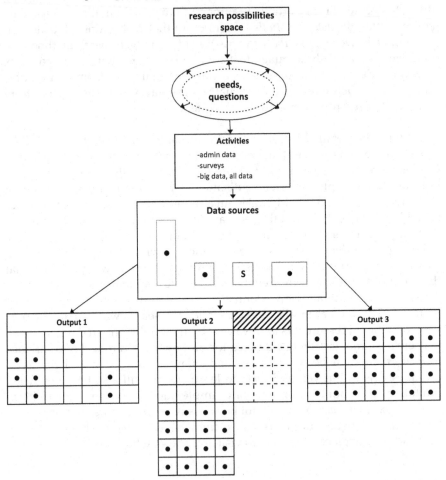

transition was already underway. With the new data sources now in the mix, the race to substitute administrative for survey data expanded to include all data. Nothing is off reach, certainly not at the outset. Regardless of actual access, the mere knowledge of the existence of the new data leads to a pause. As we design new statistical products, can we really forge ahead and daringly measure something that exists somewhere and someone knows? In the very least, can't we obtain some aggregates or benchmarks for fail-safe? Additional risks are assumed merely from the awareness that impactful developments leading to new data may also be afoot. The number of available boxes grows not only in the actual but the potential sense too – indicated by data sources with dots in Figure 7.1. It's gotten crowded out there, exciting but also dangerous. How do we separate promise from hype? A lot of money, effort, and reputations are at stake. A real-world example will start us off.

## The Humpty-Dumpty

The arrival of the Internet in the early 1990s triggered huge demands for new data. Tracking the penetration and use of the new technology was crucial to governments and businesses, and highly prized by researchers and the media. The speed of adoption would determine the timing of new ways to interact with people as clients or citizens, including the pace towards e-government, the deployment of business applications such as e-commerce, and so much more. Surveys to produce *by design* key measurements of Internet connectivity at the household level, as well as understand the frequency, intensity, type of use, and privacy concerns of individuals, began in North America in the mid-to-late 1990s. By the end of the first decade of the new century, they were conducted in many countries. They were even legislated in the EU. Bear in mind that in the early days connectivity was of the dial-up type and the access device was almost exclusively the desktop. Questions on broadband were added when always-on access arrived. These surveys produced a lot of useful data and made a difference in the design and implementation of policies and business strategies.

Two decades later, though, was a good time for sober second thoughts. Much had changed in the intervening years, including the arrival of the new data. Valiant efforts were made to explore options of using the Internet to measure the Internet. Network and user-centric approaches were investigated, experiments were made. All told, the answers were really along the lines of Output 2 in Figure 7.1. New data can fill parts of the desired output and produce extra outputs (both concentrated as dotted cells at the bottom part), but some survey data cannot be replaced (empty cells on the top part). It so happened, the latter were the crux of the matter! The number and proportion of connected households and their characteristics.

The same happened at the international level. The ITU looked at tapping new sources to patch gaps in household Internet access series due to the absence of surveys in many countries. The matter was taken on by expert groups at the UN too, among other pieces of work on *big data*, and pilot projects were conducted in several countries. They returned the same basic findings – Output 2. Lots of relevant and useful data are on tap but not the household connectivity figures.[10] But why?

The new sources are generally superior in collecting usage rather than connectivity data. Usage data from surveys relied on an estimate by an individual of the total time spent online from work, home, internet café, or any location. By now, many households have Wi-Fi and multiple access devices, and aggregation over devices and locations is not an easy walk. However, usage data are logged on servers or call detail records. Connectivity, on the other hand, relies more on subscriber data. Whether

ISPs cannot distinguish sufficiently between residential and business accounts, are unwilling or unable to share, key indicators aren't forthcoming. In many developing countries, wireless carriers also double as the ISPs for people whose only connection is through the smartphone. They may not even have good client records with addresses, as the dominant model is prepaid SIM cards.

Experiences such as those capture the predicament we find ourselves in, a clash between what we're still doing and what is now doable – which is further complicated by the fact that something feasible may be way ahead of its time practically if the data that exist cannot be had. Is a meteorite strike necessary for 'dinosaurs' in our midst to become extinct? Or, we'd all rather put on our problem-solving hats? In principle, doing away with the surveys described above is generally doable. Ignoring issues of access to the data for now, ISPs in most places know their clients. Mine easily remembers my name, my address, and the type of my connection. It has never missed to send me a monthly bill. Having microdata on residential connections from all ISPs will make it possible to arrive at a census-like national list. The type of connection would be even better reported, as some respondents might not provide the best answers when the kids aren't home. From there, we can also get the other characteristics of interest. While the account number my ISP assigns to me may be useless to the regulator or the statistical office, everything else is quite relevant. My name and address are both possible keys to link to other information about me in a population or address register. They can get my other particulars, income and family type, in a way sufficient for the intended use, which is grouping in bands.

If the needed data for both connectivity and usage components can be had, what of the couple of questions on privacy and security? Surveys softly asked if people had any such concerns. *"Oh yes"* answers were predictable, unimpressive, and never jived with observed behaviour. Arguably, on the aggregate, such sentiments can also be gleaned online from our social media behaviour. This makes privacy protection advocates cringe, as they're puzzled, irritated, and would like us to behave differently there.

Still, establishing that something is doable doesn't necessarily lead to the conclusion that it should be done – here and now. Many pros and cons need to be weighed. In our example, data from all ISPs would be needed. Missing one would spoil the approach. These data need integration through linkages with other databases, and subsequent extractions. Data on usage require many additional sources and complex aggregation. Throwing social media to the mix adds to the package. In such exercises, there may be instances where the number of sources becomes unwieldy. Not only each source fills a smaller part of the output matrix but even individual cells may need multiple sources. The information-to-data ratio shrinks significantly, with all that entails. Rather than neatly constructing an output from available data, it may look more like putting humpty-dumpty together again. If the pendulum swings too far, we run the risk of overdesign. Judgement calls are in order.

There are many practical issues too that have to be dealt with, including the modalities of accessing private data. Deliberations, pilots, and actual collaborations have started in many jurisdictions. Meantime, if there's something we care a lot about, here and now, not only do we throw our questions in a survey but we include them in the national census too. Australia did exactly that for use of the Internet questions in the 2006, 2011, and 2016 censuses, dropping them from the 2021 census when penetration was deemed sufficiently high.

The sure thing is that this new model, where all kinds of data are funnelled in, curated, maintained and used on demand, has staked its flag deep in the new research grounds. Remaining questions are of the *how far* and *how fast* type. Volunteers will fare better than those who may have to be dragged along.

## More Art than Science

An immediate requirement for any research design from now on is knowledge of a large number of data sources and their contents, from survey and administrative sources to big and all data. But the space is vast. Even a single survey has enough depth to bury someone for a long while. Understanding the administrative forms we encountered, diverse private sector data holdings, and data from social media can be a full-time job. No one can possibly know it all and fill that role. Perhaps, this is what's meant by the oft-quoted expression "*statistics is a team sport*". At a minimum, though, we need to know where to look, where the doors are. When the time comes we can drill into more details on a need basis.

Even that is not enough. We also need to know what's brewing. We've already seen more than enough to know that someone in a basement or a garage is up to something disruptive. While no one is expected to know all that either, surely we can be tuned into what major players are up to. This too has implications. A heads-up on a new solution under development could mean scrapping an elaborate design-in-the-making and perhaps joining forces through co-investment to speed things up. We've already started to see collaborations among unlikely friends.

However, skills on actual and potential data sources and their contents are in short supply. They're not included in the 'data science' skills advertised heavily since the beginning of the century. Most people going about their daily statistical business go as far as their known thematic boundaries. No one is taught broadly, and no one teaches. Universities don't, governments don't even for bureaucrats designing evidence-based policies, business don't, even statistical offices hardly scratch the surface internally. This type of knowledge is not a matter of creating and memorizing a catalogue of the old telephone book type. Wide and repeated exposure is needed, networking, and a special knack and affinity for such things – an extra dimension to numeracy. Nowadays, unintentional yet unforced errors can lead to wasteful activities. Reinventing the wheel and finding out after the fact is not part of any design.

It's also imperative that research design is grounded in reality. Practical knowledge of the nitty-gritty involved in the acquisition of data, complete with timelines and costs, cannot be underplayed. Oftentimes, this will separate a realistic and successful design from fiction and wishful thinking. For example, an important project cannot hang on the hope that a critical data component will actually materialize in a month, if it's attempted for the first time ever and requires negotiations with 50 states for the acquisition of the data. Knowing that transactions data exist in banks doesn't mean they can be had and used within the next few months of a project. Knowing that data can be produced from satellite images doesn't mean they're worth the investment (as we saw in the silly example of measuring a few backyards in Chapter 2). This type of overdesign can be akin to wasting a cannonball on a fly. But, we must equally be aware that things can change on a dime. For all we know, soon enough an app inside the camera of our smartphones will provide exact measurements on all pics, including the distance between eye pupils. Then, what appeared to be a cannonball against a fly turns out to be something quite smart actually, leading to praise for foresight. All that can be a bit messy, but that's how it is.

The inescapable conclusion is that design solutions will come from the interplay among all factors discussed and the abilities of those involved. Short term, they will even have to strike a realistic balance between a convoluted mix of too many sources and the freedom (or boredom) of a stand-alone survey. In that sense, research design cannot be approached as a right or wrong answer. Key factors will be the specifications of the needs and the intended use of the data, the actual and potential availability of sources, and estimates of time and costs. So, how could someone approach a new design? The vignette below helps demonstrate some of what we discussed.

## A Morning Message

As she's having her morning coffee, a colleague working at a statistical outfit gets a text message. A name and a number are provided together with the message *"Your director gave me your coordinates. I'm interested in data for swimming pools. Please call later today"*. A bit of an unusual request, but in her line of business she's had stranger ones. She'll call him from the office. She's good with customer service. She also knows that her outfit has no data on swimming pools and she can't think off-hand of any source that would. But her boss would know that too, he wouldn't have referred him to her for no good reason. Who could that person be, what exactly does he want, and why? What if it's not an availability request, what if she's asked to produce such data through a new cost-recovery project? Now, she's hooked. On the way to work, her creative juices start flowing. How could she do it? She mentally separates residential and commercial pools. Images of pool builders, manufacturers, and importers cross her mind but only for a fleeting moment. She discards the supply chain. If a quick ballpark national estimate is needed,

**FIGURE 7.2**

Alternative designs for statistical outputs

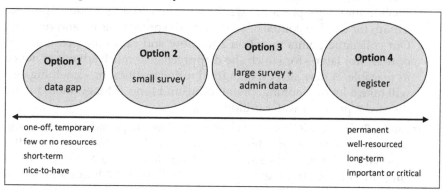

a household survey would do. Rather than a new one, piggyback with a question or two somewhere – perhaps even the labour force survey. If she pushes enough it won't take too long, it could even produce data by province. But what if he wanted more detail – data by city or neighbourhood, by type, by year, and who knows what else? Would he have the money to spend and the time to wait for a very large and dedicated survey? Should she explore the option of municipal permits? Being still some distance from the office, and in the spirit of exploring options from the Yaris to the Cadillac, she keeps at it. What if he wants to track swimming pools every year? She definitely has to ask if it's a one-off. She couldn't possibly repeat a survey and ask the same people the same question! She dares to imagine the construction of a complete inventory, a register that could be updated annually. This could be really big, any source is at play. Other than building permits, data from satellite images and other possibilities come to her mind too. Before she gets completely ahead of herself, she's arrived. At the office now, she makes the call. Her thinking, in conjunction with the scenarios below, gives rise to Figure 7.2.

> **Scenario 1:** A journalist is on the other end of the line, interested in swimming pool data for a story he's working on. It'd be nice to have them. Our colleague explains politely that this is a data gap and apologizes for not knowing any other source. She points to the very left of Figure 7.2. The same would have been the case if a student working on a term paper was on the other end of the call. Bummer!

> **Scenario 2:** The person on the other end is an official from the national association of water utilities. They're at the early stage of research on water consumption specifically from swimming pools. Many of their members are seriously considering a two-tiered pricing scheme, supported by water meters dedicated to swimming pools. They need something quick for now, for rough calculations of water volumes.

A ballpark would be fine. If this work gets any traction, though, soon they'll need good quality data for the number and size of swimming pools in the geographical areas served by their members (they'll provide a map) for another couple of years until the program is up and running. Our colleague points to option 2 for now and raises the possibility of option 3 for later – for which she'd appreciate as much advanced notice as possible. A new survey, data from municipalities, or something else will take a lot of planning, preparation, and time. (Money shouldn't be the sticky point for water utilities.)

**Scenario 3:** The call is from the chief of staff of the president's office, the minister of finance, or whichever authority has jurisdiction over the following. The government will roll out a new energy initiative in support of its environmental policies. It will be fully funded by a new tax on the super wealthy and a few other measures, including a surtax on all swimming pools, except those municipally owned. Rather than allow residential or corporate swimming pool owners to self-declare, the decision is that they'll all get a bill. The immediate requirement is a complete list of addresses for residential and commercial establishments to set a rate in time for the next budget. Going forward, the requirement is a permanent and detailed database, updated on an annual basis in support of the new program. Ideally, it should contain information that will eventually support the refinement of surtax options based on the dimension, age, and value of the pools as well as the value of the properties and perhaps other characteristics not yet decided. Money is not an issue. Our colleague points to option 4. She also mentions that her statistical outfit can definitely help.

Moving from the left to the right of Figure 7.2, the design morphs. It depends crucially on the nature of the need, its intended use, and who's asking – not as in the clout they have but in the resources they're willing to put behind the request. It's not that our colleague didn't care for the journalist or the student, they're data users after all. She'd gladly use her expertise and suggest ways to help them out with existing data. But what is she supposed to do with all the requests she receives if they involve 'data gaps'? Regardless of the spectrum of options, a data gap can be patched temporarily or can be closed permanently. If and when a new register is built, Figure 7.2 vanishes. Once up and running, it will satisfy the needs of the president's office, the association of water utilities, the journalist, or the student equally well. This is a familiar junction. Linking to our previous exposition, the more the data come alive, the more the possibilities space expands. This expansion manifests itself also in more requests for data in gap areas, which explains the seeming paradox that even having more data than ever there are also more unsatisfied requests and complaints for a lack of data. While many gaps of yesterday have been taken care of, many more take their place. The process will continue on.

Nonetheless, our colleague's work is not quite done. A decision on *what* to do only triggers the intricacies involved in *how* to do it. This is now real. Initial thoughts for data sources, processes, timelines, and the like no longer cut it. Pragmatic planning and risk management are now in order. Options will be weighed, the pros and cons of alternatives will come into play. This is the essence of informed decision-making behind design choices today. Satellite data may be a good source for a register of swimming pools. They can provide complete stock counts and specific locations, measure dimensions, and can be relied upon for continuous updates. On the other hand, values or socio-economic characteristics cannot be had from such data. Municipal permits and other sources will be needed. The more sources are explored during the build-up phase, the better. Even if they overlap, they'll enable data confrontation useful both for data quality and learning purposes. Looking forward, subsequent linkages with other registers will also be considered. The implementation of any decisions will take time and money. Then, what if an earth observation outfit, a GIS company, or a student in a dorm is at the final stage of developing exactly the algorithm needed to convert satellite images to swimming pool data? You can't know enough. Is there room for alliances? Many things will come up, and they should. This is the time. Trade-offs will be identified and contingency plans will be drafted. With enough unknowns, it's even possible that an interim design with key milestones is favoured, leading towards a more optimal longer-term structure.

The discussion in this section offers an opportunity to also broach an issue that will become contentious in the world of the new data. We said that our colleague works in a statistical outfit, but of what kind? Is it the NSO, another government place, or a private business? *Who* will build that register of swimming pools? The abundance of fresh data sources changes the previous dynamics of data collection and tilts the balance of powers. If the backbone of the new register is satellite data, at first glance the NSO has no advantage over anyone else if they all start from scratch. The same would be true for statistical products fed by search engines, social media, or sensor data. The legal power to compel response has no sway over freely available or commercial data. Surely, the NSO could build a register and link it to other data sources too but, as things stand, it won't be able to fulfil the ultimate need and supply a mailing list. The data it produces support government policies and regulations *indirectly*, through analytical findings and enabling research in policy shops. Operating under the legal framework of a statistics act, it can't provide data that would identify individuals, households, and businesses – not even to the president's chief of staff. With legal wrangling, some way will be found, even if such work is uncharacteristically done outside the Statistics Act. With a whiff of irony, another government entity will be freer from the confidentiality clauses of the Act. It may not possess the necessary statistical expertise for such an undertaking but it could solicit competent help for certain tasks, including from the NSO. A private business receiving the call for such a project would hardly hesitate to take on the challenge – and a lucrative contract. Free from legal obligations, it can serve the president's office with its mailing list. However, it'll be a bumpy and challenging road to get to it – not to mention the need for linkages at the end. Still, an argument can be made that the identity of the ultimate client would help! One way or the other, building a register is a societal investment in data infrastructure. In parallel with the effort, the time, and the expense we must grapple with confidentiality concerns, which won't just go away. Decisions must be made on how others will have access, what parts of it, and under what conditions. Are there parts that can be made public? Addresses are already public. Would the addition of an ostentatious swimming pool be confidential?

## Data Flows and Intercepts

Data are born somewhere, then they circulate widely. Where best to find them was a matter lurking in the background in the world of surveys but it's front and centre in the new world of data. There's always been a level of awareness that a data need can be met with demand- or supply-side data holdings. The survey instrument would be directed to the better source, appropriately. If an estimate of total cigarette smoking nationally is needed, a few tobacco manufacturers know their sales well. If a health agency cares about heavy-smoking neighbourhoods, estimates of sales from local corner stores may be preferable. If the proportion of smokers is at issue, a national survey is needed. Efforts were made to make sure that those surveyed could provide the answers.

Looking through the prism of a *systems approach* provides insights on what has changed. A classic *network* shows actors/agents (nodes) engaging in a multitude of interactions (edges). The nodes in Figure 7.3 represent individuals, households, businesses, governments, buildings, and more. Nodes are very conducive to turn into full-blown registers. Moreover, the nodes shown are really top layers. They encapsulate more detailed networks underneath, with individuals as the nodes (shown on the right part). The same magnification applies to businesses. Collections of individual units (shown in boxes) can be thought of as households, industries, or something similar. Historically, data were collected *from* and *about* these nodes, without asking one about another.

There's been some criticism that too much emphasis has been placed on nodes at the expense of the edges. But this isn't quite so. It's true that an astonishing number of questions has been asked over the decades, particularly of individuals and businesses. A laundry list of questions like *"who are you, how old are you, how much do you have, how much do you want, what are you up to"* were used to *profile* a node. But a long line of questions has also chased after the edges, the *data flows* that record interactions, transactions,

**FIGURE 7.3**

A classic network with actors/agents (nodes) engaging in a multitude of interactions (edges)

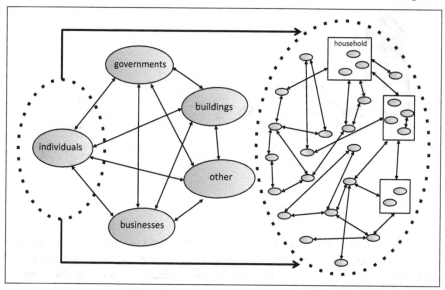

and relationships among nodes. At the end, though, all edges land on the nodes. That's where the answers are held.

The relationship between nodes and information holdings has now changed. In the past, having paid with cash for a TV set, a buyer walking out of a store could answer the question *"do you own a TV set?"* The seller's record enabled him to answer only that he sold a TV set but not to whom. With electronic payments, sellers know the identities of the buyers – especially online. Supplying companies, other retailers, or manufacturers may also know. Financial institutions are able to identify both the buyer and the seller. And let's not even follow the trail of other intermediaries, such as the post office or delivery companies. A transaction that generated a single record in the past generates a whole bunch now. Think of all those who know, and what they know, of your latest Amazon delivery.

First, as recorded observations, data generation and flows have increased and this is one of the reasons why we have more data now. Multiple 'aspects' of the same transaction are recorded. This is consistent with developments elsewhere, such as having a bigger part of our lives documented and recorded. From perhaps a sketch to capture a place, a face or a moment in much older times, to a black-and-white photograph later, to multiple video angles of the same goal scored today. The struggle of the previous generation to find a childhood photo has been replaced by the struggle to choose from dozens of photos in a minute. The second implication relates to the identification of the

**FIGURE 7.4**
Where best to intercept data

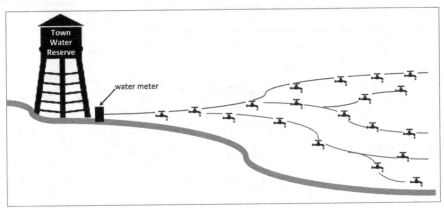

buyer, which is now circulating with the data – and as data. As a result, the capacity of the supply side to provide data grew, offering more choices. Legal and ethical matters aside, from a purely statistical perspective, there are at least two nodes that can give an account of the same record.

Once again, where to intercept data depends on their intended use. Domestic manufacturers and importers of TVs do have records. Aggregating across all of them, and over decades, will produce some total of the stock of TVs in the country – if that's what we're interested in. This may show that there are more TV sets than households in the country, whereas a household survey could indicate that half the households don't have any. Both data points may be correct, if we account for every hotel and motel room with a TV set and all those thrown out by now. If we're after extreme accuracy, we can always ask for registration! (We saw that this has been attempted.) These days, data from financial transactions are also poised as good candidates to hold such answers. There are alternative sources for the same data. The choice of where to intercept won't be dictated solely by intelligence related to their flows but also by jurisdictional, legal, cultural, and ethical considerations. However, in a very simple system (Figure 7.4), when the volume of water used in a town over a period is needed, the meter at the town's water reservoir will beat measurements at individual taps every time.

Hard data from nodes came from memories and records, and respondent recall was always a problem to reckon with. To preserve memories of special moments, there was Polaroid. Now, if you ask me for the places I've been during the last year or what I bought, I'll have to check with Google and Amazon if I care to answer well. Individuals are at a disadvantage even when it comes to hard data, but there's a Polaroid moment everywhere and at all times now. Our hard data are strewn all around. If we're okay with so many having them, who else can we give consent to? Whose data are they, anyway? As for softer data, they refer mostly to plans, intentions, sentiments, and opinions. They were always subject to swings, and put to more ephemeral uses. Much of that is also online now. So, if our hard data moved out on their own and our soft data can be gleaned online, what else is left for anyone to ask from this particular node? Our very inner thoughts? The real ones or those on social media?

---

## The Graduate Course

The setting is a lounge that doubles as a classroom for seminars. A professor walks in, nods to the few graduate students gathered around, and hands out a sheet with a list of research topics for the semester. He proceeds to say "*This seminar is about the pursuit of knowledge, free from biases or preconceived ideas. Crossing boundaries is expected. You must push aside everything you've learned up to now, to clear space in your mind. What you think you know from all your past As and B+s are half-truths at best. Ask good questions and search relentlessly for answers. If you find any answer don't believe it, torture it. But if you believe it, please tell me – I'd like to know too. Oh, by the way, there are no office hours.*" Whether you've been in such a setting or not, there comes a time that all knowledge needs to be moderated by some lingering doubt. Once in a while we need to turn things we take for granted, together with the comfort of our routines and our stereotypes, into footstools, stand on them and take a peek higher in the horizon, ever so humbly lest the brightness of the sun blinds us. How do we do this? I'll give it a shot in the remaining few paragraphs.

The *survey* offers a good opening. We've used it, overused it, loved it, loathed it. We're all very familiar with it, though, right? But what exactly is a survey? Stereotypically, the image of a few pages with questions and tick boxes comes to mind, perhaps with a cover page informing us that this is a survey, instructing us to answer truthfully and to the best of our abilities, and assuring us that our responses will be kept strictly confidential. If it comes from an official source it may also include a toothless threat for nonresponse. There have been surveys with tens of pages, rivalling printouts from old mainframes that made people's knees buckle. On the flip side, there have also been surveys with one single question. There have been surveys that could be answered offhandedly and others that required digging into archives, accounting ledgers or keeping logs and daily diaries.

But what are the pages of the *application form* for a passport? What are the questions and the boxes for answers in the forms required to renew license plates or open a bank account? And what about the income tax returns every year, with all their 'schedules'? Each one of them is much more onerous than the census. What would change if someone stuck a cover page upfront all of the above and asked us to fill the 'survey'? What if you call someone at a municipal office and ask him to email you a list with all the arenas in town? (To which he'll reply that they're online.) What if we 'survey' a city for its administrative records on marriage certificates or a utility for its smart meter data? What of researching the price of a couple of gadgets in a few manufacturers' websites for some price index? You can think of countless examples, even including questions we ask of our friends and family. The point is that we shouldn't be hung up on the idea of a survey as we (thought that we) knew it but get past it – and over it. It doesn't matter what label we stick on a request for data – a survey, an application form, a web search, a telephone call, or merely a question. What it boils down to is that at some point someone is asking someone else for some data through some means. This is the life and fate of data. Recall our discussion of nodes, data flows, and points of intercept.

A more dicey matter concerns *data linkages*. There is a certain ambivalence about them, which leads to unease. Some swear by data linkages as ingenious and powerful shortcuts to new data and knowledge. For others, linkages have a bad wrap as they raise confidentiality and privacy concerns. The more we link, the greater the risk of revealing sensitive information that can identify individuals or businesses. In that context, even within statistical shops, data linkage activities are subject to stringent procedures with justifications, approvals, and retention periods. But looking higher up, where the sun is brighter, there are knots that need untangling. In the very least, preconceived ideas hailing back to the old world of data need dusting. Let's follow along a little.

Say that our research focuses on a particular group of individuals. We ask each of them (a survey?) *"how old are you?"* With no inhibition whatsoever, they all respond. We now know each person's age but, to protect their private information, we make public a frequency distribution like the top-left panel of Figure 7.5. It takes no time at all for the interest to shift to their income. We repeat the survey exercise asking *"how much did you make last year?"* and end up with the top-right panel of Figure 7.5. We now have more knowledge about this group of individuals. However, rather than celebrations our accomplishment triggers arguments – such is the nature of research. Someone argues that young people are loaded today, they make more money than older individuals. Someone else, still remembering the lifetime hypothesis from his economics course, forcefully maintains that this can't be so, older people earn more money. By now each of our two data sets has exhausted its information-generating potential, and we can't go back to the group a third time as no individual there can directly provide such an answer. But we can get to the truth by putting the data we have next to each other (data linkage). We thus arrive at the bottom panel of Figure 7.5. Using both the average and median incomes, it shows that in this particular group income increases with age up to the 45–54 age group before it decreases. That age group makes more money than any other. Also, the income of the 25–34 age group is higher than that of the over-65 group.

**FIGURE 7.5**

Population by age, by income, and income by age

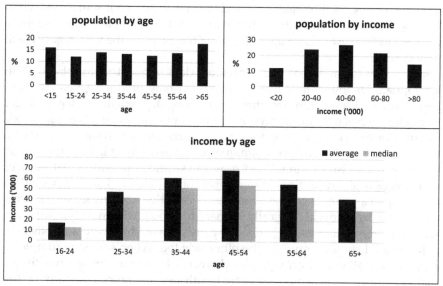

Salient points that need to sink in here are:

- Conceptually, it makes no difference whether the two data sets are obtained from two different surveys or a single one, by the same organization or different ones.
- The new knowledge is generated solely by merging the data points we already had. This is the power of linkages. Those who provided the data didn't have any more to give.

In the spirit of the previous discussion, we can combine *what's a survey* with *what exactly constitutes a data linkage*? Had we asked both questions on age and income in the first survey, we wouldn't be talking about any linkage. We, or someone else, can easily start all over with a 'unified' survey asking both questions and many more. Clearly, that would be wasteful since we have all that's needed to answer the new research question. The lack of foresight to think of a particular need ahead of time should not become a penalty either for the researcher or the respondents. Even if we had thought of that before-hand, there will always be something else, a 'next question'. When the same population is concerned, we can break a data collection instrument into as many pieces as we like. In the limit, we can perceive all surveys in a year as one gigantic survey with many tentacles (modules). Respondents can be informed that they've been selected for a survey which will be administered in 37 individual pieces throughout the year, to ease the burden of the imposition. We can then 'link' all the data without any 'linkage'. We can even go about business-as-usual if we wanted, with different but overlapping samples across surveys for limited linkages. Methodologists would find a way around it. Our introduction letter would simply go to the combined populations sampled, rephrasing the message to *"you may receive up to 37 parts of the survey"* – the vast majority will only get one.

In the new environment, our individual data frequent many places. With appropriate identifiers they can be linked, not only producing new data but potentially saving us from irritating, groundhog-day repetition. Trying to get through to a business these days, someone may have to give the same basic information two or three times during one phone call – when she's not on hold, that is. Remember my half-joking unique individual number in Chapter 3? Well, the other half wasn't joking. What if, whoever asks, I provided that number which leads to all my particulars somewhere on the 'cloud'? I, for one, would go for it – up to a point, of course. From this perspective, the privacy argument is weak, if not hypocritical. Logically, it's hard to fathom what data that we share with an online retailer or a delivery company we wouldn't share with others.

To be fair, I should argue for the other side too. When someone completes a purchase online and arranges for delivery she neither knows nor thinks that her information would be shared with her insurance company or the statistical office (regardless that the latter already possesses her tax data – not to mention it can hit her with a mandatory survey anytime). In fact, she's more likely to receive assurances that her information will be kept 'strictly confidential' – something that may not be backed by any legal power and may not happen.[†] The argument that personal data cannot be shared behind someone's back is a legitimate one. What if, instead, she was informed that some 'basic' information will be shared widely? Say, if she consents? Would that stop her from proceeding with that purchase? Or she belongs to the large category of people complaining why they're asked repeatedly for the same thing, particularly from their government?

This nut is tough to crack, no easy answer. A naysayer will still say nay and will certainly demand a precise definition of what's included in 'basic' information. Name and address is one thing. Age and gender another. Google offers to autofill your credit card information too. There are some very private health issues and other sensitive matters. After all, we don't want an insurance company barging into people's homes, catching them eating french fries and cutting them off. (This has allegedly happened in the US.) Moreover, even consent is not a panacea. Linkages work best with complete and census-like data sets. Linking across incomplete sets or surveys with different samples, for instance, the overlap may be too small and useless for inference. But that's

---

[†]   In the early 1990s I subscribed to a monthly magazine. In the process, not only they misspelled my name but they completely bungled it. There were two typos in my last name (I understand) and another two in my first (harder to fathom). For about 25 years I was receiving all sorts of unsolicited mail with that spelling.

exactly the sort of thing we ought to be talking about more, to figure out how best to move on in the new world of data. The statistical and analytical usefulness of data linkages should be de-coupled from the need to safeguard confidentiality and privacy in public releases. The latter relates more to aggregation, anonymization of data, masking procedures, application of proportionality frameworks, and a host of other mitigation strategies related to the dissemination of public outputs. What if, instead of statistical outputs, matters of life-and-death were at issue? Is this an area, perhaps, where we need more awareness of the issues and a change in culture and attitudes more than the law? Or has this already started?

## Fact-Checking Tips

*The Google Flu Debacle:* When something new and promising appears on the horizon, hype is not far behind. And nothing sustains hype better than a powerful demonstration, non-believers want to see miracles. Early research with Google data showed promise and attracted press, good for a new enterprise. Emboldened, Google took on a bigger challenge: outdo the CDC in quantifying flu prevalence. The flu inflicts millions every year and causes death. Information on local outbreaks is invaluable to prevent or contain further spread. The earlier we know, the better off we are. Starting in 2008, and with an eye on the CDC's flu tracking information, Google claimed that its nowcasting algorithm, Flu Trends, could produce accurate estimates of flu prevalence by aggregating over millions of queries. It could best the CDC by up to ten days. Turning traces left behind by keyboard strokes into potentially life-saving insights is a big deal, and looked to be working for a while. Until cracks started to appear, then widened, and ultimately the bubble popped.

The algorithm failed spectacularly. Among a series of anomalies that impacted its credibility, it was off by 140% at the peak of the 2013 flu season. As explained by a paper in *Science*,[11] Google's algorithm was *"vulnerable to overfitting to seasonal terms unrelated to the flu"*. Just because someone types fever, cough, or running nose doesn't necessarily mean they're searching for the flu. On the other hand, any number of terms that aren't flu symptoms may well be used to look up flu information. Criticism included that Flu Trends didn't factor in changes in search behaviour over time. Google quietly euthanized the program in 2015. As the *Science* article stated, the poster child of big data turned into the poster child of the foibles of big data. However, this particular failure doesn't wipe out the value of the new data as much as it highlights problematic uses.

Knowing the nature of data, and their known caveats, their use can avoid overreach that risks misinterpretation. To stay within bounds, it matters whether searches took place because of something of interest or unrelated to it. That 15,000 fans congregate at the Montreal Forum is no news when a Canadiens hockey game is on. If it happens for no known reason at all, it's another story. Online chatter for a bombing that hasn't happened isn't just some buzz but one for the authorities. If rumours of an unidentified flying object (UFO) sighting circulated in Ottawa, recording half-a-million searches for extraterrestrial life has its meaning – which doesn't extend to that half-a-million individuals believe in UFOs. It'd be nice to know if there were also half-a-million searches for the Ottawa Senators that same day. *"Where there's smoke, there's fire"* is absolutely true. Let's not also forget that smoke signals have been used for communications. Tracking smoke to infer fire makes sense, and many times this is what happens with search engine data. But winds will spread the smoke far and wide, weakening its connection with the fire – we should be able to tell. From the smoke, we can't assess the extent of the damage caused by the fire, and certainly we can't extinguish the fire by putting out the smoke.

We've maintained that there is value to such data. There's a time and a place. There may also be an approach, perhaps combining them with other sources in research design as we saw in this chapter. Data from search engines have many converts, they're now used extensively and experiences accumulate. In the book *Everybody Lies*, Seth Stephens-Davidowitz maintains that people are liars at heart – but we tell Google the truth in our private moments, like in a confession booth. There's truth to the lies! Survey takers knew all along that you don't ask people certain things, and if you do you don't quite believe the answers. We're mostly pious, and it's hard to find anyone who knows how to spell porn or bigotry. But moving from there to Google as the serum of truth is a very long stretch, there's a lot of middle ground. I won't speak for you but I admit that half the times I google I settle for one of the autocomplete options that come up fast. This feature turns the search engine into a mind reader. One can argue that not only my searches don't represent the truth, but they're not even mine – they were suggested to me. The results would reflect the power of those suggestions. I won't even bother you with the other half of my searches – but some are genuine. So, if you ever obtain such data on identifiable individuals (something unlikely), and your research depends on accurate data for each individual (something undesirable), you'd be better off to remove mine. In addition to the discussion here, a frequent fact-checking advice on how to spot *"big data hubris"* is *"if it's too good to be true, it probably isn't true"*. Yet, again, cold fusion may still come – someday.

*Facebook woes:* Developers were allowed to create apps for data collection purposes on the social media platform. To build psychographic profiles of voters, Cambridge Analytica combined such an app with a quiz to collect data from consenting users. However, behind Facebook's back, the app explored a technical loophole and collected the data of their 'friends' too. This compromised the personal details of tens of millions when the data were sold to political campaigns, including the 2016 US presidential election.[‡] While this had become known for some time it was publicized widely in 2018 thanks to a whistleblower, leading to the demise of that company. Facebook settled for hefty fines. A lesser known story is that shortly afterwards Facebook entered a partnership with *Social Science One*, a research consortium of academics whose mission statement laments: "*Not long ago, academics had access to almost all the research data in the world because we created it or easily obtained it from others. Today, the most informative data about people and societies is collected by private firms that offer no provisions for academic access.*"[12] The consortium would facilitate the challenges of negotiating data access with corporate attorneys, communications teams, program managers, project leaders and the like, which are such that "*…merely answering the phone can be overwhelming*" even for a research centre, let alone a single individual researcher. For Facebook, this was an opportunity to atone by showcasing a new 'model collaboration'.

In February 2020, one of the biggest data deliveries in the history of social sciences took place when anonymized data, allegedly containing 42 trillion numbers, were delivered to the consortium to study the effects of social media on elections and democracy. Members of *Social Science One* got busy. A year and a half later, Fabio Giglietto from the University of Urbino in Italy was doing some fact-checking – sort of. Comparing these data with a Facebook report released in August 2021, he detected anomalies which he reported to Facebook. If the researchers had 'richer' data, they would have discovered the error earlier – others said. Facebook looked into it and by September the news broke that the data were indeed faulty. Due to a 'technical error' the US data set missed about half the US users. It so happened, this half was those who are not so politically polarized and their leanings are not detectable – talking about a bias given the objectives of these studies! At the time of writing, the damage is unknown but data were shared with at least 110 researchers who published, or are on the way of publishing, an unknown number of papers.[13]

For fact-checking purposes, the lesson is twofold. First, the way the error was found is classic. A competent data user 'feels' something is amiss and notifies the data producer. An honest data producer digs deep, finds the problem, and implements mitigating actions. The error might have been due to a 'technical' matter, an issue around data 'hand-offs', a new and inexperienced employee. Second, a good researcher typically compares findings across different pieces of research using similar data. In a subject matter area you know, you can spot a poisoned glass. But what do you do if all glasses are filled from the same well, and the well is poisoned? One of the academics

---

[‡]  More recently, Facebook 'de-platformed' other researchers collecting data in a similar manner. Browser plug-ins that scoop up data from consenting volunteers on how ads are fed to them is something that Facebook doesn't like, doesn't make known, and considers unauthorized scraping – citing privacy concerns.

involved was quick to ask for government regulation to force social media companies to develop secure data sharing infrastructure with researchers. Hold your thoughts until Chapter 10.

## Notes

1 Interesting anecdotes are contained in David Salsburg, *The Lady Tasting Tea: How Statistics Revolutionized Science in the Twentieth Century*. Owl Books, 2002.
2 Some well-written and thought-provoking narratives can be found at: Martin Hilbert, "Social Media Distancing: An Opportunity to Debug Our Relationship with our Algorithms," 2020, https://martinhilbert.medium.com/social-media-distancing-an-opportunity-to-debug-our-relationship-with-our-algorithms-a64889c0b1fc
3 Facebook, Data Policy, www.facebook.com/privacy/explanation/ (Date of Last Revision: January 4, 2022).
4 International Telecommunication Union (ITU), *Measuring digital development: Facts and figures 2019*, www.itu.int/en/ITU-D/Statistics/Documents/facts/FactsFigures2019_r1.pdf
5 Twitter Inc., Twitter Privacy Policy, https://twitter.com/en/privacy/previous/version_14
6 Quoted from: Tim Davies, Stephen B. Walker, Mor Rubinstein and Fernando Perini (Eds.), "The State of Open Data: Histories and Horizons." *Cape Town and Ottawa: African Minds and International Development Research Centre*, 2019, www.idrc.ca/en/book/state-open-data-histories-and-horizons
7 Several similar and additional issues are discussed in more detail by Derek Ruths and Jürgen Pfeffer, "Social Media for Large Studies of Behavior: Large-Scale Studies of Human Behavior in Social Media Need to Be Held to Higher Methodological Standards," *Science*, November 2014.
8 From October 2018 to March 2019, Facebook said it deleted 3.39 billion fake accounts, more than its 2.37 billion active users at that time. Niall McCarthy, "Facebook Deleted More Than 2 Billion Fake Accounts in the First Quarter of the Year," Forbes.com, May 24, 2019.
9 Cornelia Brantner and Jürgen Pfeffer, "Content analysis of Twitter: Big data, big studies," Preprint of book chapter: Brantner, C., & Pfeffer, J. (2018), in S. A. Eldridge II & B. Franklin (Eds.), *The Routledge Handbook to Developments in Digital Journalism Studies*. Abingdon: Routledge.
10 International Telecommunication Union (ITU), *Big Data for Measuring the Information Society: Methodology*, Geneva 2017, www.itu.int/en/ITU-D/Statistics/Documents/statistics/Methodological%20Guide%20and%20Proposed%20ICT%20Indicators%20Based%20on%20Big%20Data_27Feb2019.pdf
    United Nations, *Handbook on the Use of Mobile Phone Data for Official Statistics*, UN Global Working Group on Big Data for Official Statistics, September 2019, https://unstats.un.org/bigdata/task-teams/mobile-phone/MPD%20Handbook%2020191004.pdf

UN, Committee of Experts on Big Data and Data Science for Official Statistics (UN-CEBD), *About Mobile Phone Data*, 2021, https://unstats.un.org/bigdata/task-teams/mobile-phone/index.cshtml

11  David Lazer, Ryan Kennedy, Gary King, and Alessandro Vespignani, "The Parable of Google Flu: Traps in Big Data Analysis," *Science* 343 (14 March 2014), pp. 1203–1205, https://gking.harvard.edu/publications/parable-google-flu%C2%A0traps-big-data-analysis

12  Harvard University, *Social Science One*, "Our Mission," https://socialscience.one/our-mission

13  Article in Washington Post by Craig Timberg, "Facebook made big mistake in data it provided to researchers, undermining academic work," *Technology, Washington Post*, September 10, 2021, www.washingtonpost.com/technology/2021/09/10/facebook-error-data-social-scientists/

# 8

---

## It's All About the Microdata

---

*"You can't see the forest for the trees"* is a common saying. It's meant to deliver a powerful punch that brings closure to arguments questioning the superiority of the macro. Who wants to miss the *big picture* after all? Yet, deep down, scientists and researchers don't really like 'closure'. They sense benefits in keeping things open – remember that lingering doubt? So, respectfully and humbly, they enjoy a good argument. No need for loud voices, name-calling, or insults – just the tricks of the trade. While macro vs. micro is debated in many disciplines, economists do it best. Below is my version of the duel between a microeconomist and a macroeconomist.

---

### Macro vs. Micro

The macroeconomist's forest-for-the-trees argument always delivers a strong blow. It can be disarming and, against a lesser opponent, it can even win the day. Its impact hurled the microeconomist against the ropes. He's down – but not out. He has a few tricks under his sleeves. First, he'll use the very strength of the blow he absorbed and deflect it back, turning it into a question.

*"What do you mean by the 'big picture'?"* he asks the macroeconomist. He knows well what answer is coming back, but he gets the ball rolling and buys time.

*"We're interested in the economy at large, not a single individual or business. Output growth, low unemployment, we need to tame inflation, not the price of milk. It's the whole package that really matters, the micro is just noise, don't you see?"* As this was delivered, the microeconomist gains some distance from the ropes and picks a few more items from his stash.

*"Are you implying that the total is bigger than the sum of its parts?"* he asks, and immediately doubles down with *"How many mediocre players can a team have and still be great?"* For good measure, he even throws in a random *"so what?"* – just for the heck of it.

Good answers come back from the macroeconomist. *"Microdata are only raw ingredients, it's what you do with them that matters. You need a recipe, that's what macro gives you, can't make a cake without it"*. And, without a pause, a

DOI: 10.1201/9781003330806-8

back-to-back jab. *"If you really want to know another truth, people can't handle microdata, OK? They're too messy...there, I said it".*

The ante has been raised. Breathing better now, the microeconomist decides it's time to pull a higher-level tactical move, reverse psychology.

*"I see...thanks for explaining all that, I mean it. I'm really a big-picture guy too. Big pictures always fascinated me. I'd like to make one myself someday. Since you're closer to the area, you think you can give me a hand, perhaps show me a sample of how it looks like? How do I do it?"*

To which the macroeconomist retorts: *"Yeah, we could give it a try".* In a markedly conciliatory tone, he looks down and goes on to say: *"But first we need microdata".* Then he reaches for his backpack.

Bull's eye. The microeconomist knew that sooner or later he would force this admission. Not only he's no longer cornered but surely he tied that forest-for-the-tree argument, he's in the game.

This may all feel like a roundabout way to what seemed obvious from the get-go. We all marvel at a rainforest picture taken from above. Equally, of course, there's no such thing without the individual trees. But the macro and the micro forces are going at it for decades, including in the world of data. The catch is to get to the point that caring for microdata is not *instead* of caring for the big picture but precisely *because* of caring for the big picture. Remember the analogy with the empty lot and the thin air before the condo building stood up? The big picture is the building. People liked it, and the proof is that they bought into it. But it didn't exist. Macrodata were the blueprints, the how. The materials that built it are the microdata. The same microdata can be used for different blueprints. Talking about superiority! At this point, a very humble researcher would call it a day. But our microeconomist is only moderately humble. He now has the upper hand after all, he got an admission that we need microdata. He won't go for the jugular, of course, just after a little more devious fun.

Pulling something out of his backpack the macroeconomist says: *"Here, I have two big pictures with me. It's the same really but at different times. Have a look".*

Ever so observant the microeconomist seizes the moment instantly. *"I see that a few things are different from one to the other. Here, here, here, and there, you see? Why? How do you find the reasons for these changes? You suppose you can explain them?"*

Wow...masterful, soft-spot blow. But not under the belt, perfectly legit. Research is all about asking exactly these *why* questions and toying with the *causal forces*. Concessions kept coming now.

*"Not yet, I can't. The only way to do that is with the microdata. I need to dig in".*

Straight up! Mic dropped. But the microeconomist employs a last, tried and tested, gimmick. He recaps, slowly and clearly. For added pizzazz, he makes the words come out of the mouth of the macroeconomist.

*"So, you're saying, to arrive at the big picture we need microdata, right? And then, you're saying, with the right recipe microdata can make a cake, and the same*

*microdata can also make different cakes if we wanted. You're also saying that to explain anything that happens to the big picture, we absolutely need microdata, right? Got it, thanks...but, sorry, can you please remind me what this argument was all about?"*

Come-back complete. And to think for a moment that the powerful forest-for-the-tree argument had done him in...we'd have to skip this whole chapter. But it's here, if you're interested.

## Off to a Slow Start

There's a crew behind every personality on camera. We don't see them, the spotlight doesn't shine on them. But without them there's no show. One way or the other, microdata also underpin every known statistical indicator, famous or less so. They define the boundaries of any truth we can hope to find about a population. The whole truth is inside those census boxes we encountered earlier. How deep can we dig or how well can we use the microdata for answers is a different and longish story.

Having it all there, one record at a time, is precious but not useful. We can't just point to that collection and say *"all the answers are there, help yourself"*. So, we roll up our sleeves, draw lines, and selectively pick and arrange microdata to construct numbers that are both useful and can be understood. These can be simple counts for groups of interest, as we'd seen. How many are we, how many men and women, what's our age, where we live. Others look more like the abstract notions we've created, the GDP, the rate of unemployment, or violent crime. We're aware that some of those are received passively, while others can be hotly debated. Some rely exclusively on microdata, others less so. But until recently we rarely had the pleasure to meet microdata in public.

There are several reasons why microdata were not as prominent in the past, and why they were largely out of sight in the 20th century. One reason is that we couldn't handle them very well, the macroeconomist got that right. Microdata were never a good match for a world based on paper. Nor were they good companions for innumeracy, as they're more demanding than catchier macro indicators. Thankfully, digitization eliminated this obstacle. If anything, we've overshot the solution. *Big microdata* are so vast now that risk becoming inaccessible to many. We're past the spreadsheet, even databases, and into the world of data analytics.

Another reason why microdata didn't see much daylight earlier has been confidentiality. Individual records contain the private details of individuals and businesses, and cannot go out in public naked. This will continue, of course, but steady progress has been made on that front too. Methods and techniques to anonymize files have been devised, making many available for research. This is coupled with institutional arrangements that put in place procedures for researcher access under carefully monitored conditions. Microdata from search engines, social media, and other sources are also accessible now, as we saw. Commercial data sets have been developed as well, relying on a wide variety of public and private sources. Moreover, volumes of microdata are released as *open data*.

The dominant research paradigm with its reliance on surveys as the principal data sources did not favour microdata either. The end outputs didn't target answers at a granular level. Other times, surveys with no independent storytelling potential were also run. Their only purpose was to support the construction of some aggregate statistic, including feeding the insatiable appetite of the GDP. Being subservient to other causes, any microdata collected were sacrificed for the greater good. They were invisible supports, like underwater posts for the pier above. Moreover, statistical registers had not yet advanced past their role as survey frames. With liberated microdata, this is turning on its head.

Cultural attitudes favouring macrodata was yet another reason for the neglect of microdata. Partly fed and exacerbated by all previous reasons, such attitudes were particularly prevalent in economic statistics. Built largely in the post-WWII period, macroeconomic statistics were heavily influenced by the then-new orientation of policies, supported by related theoretical advances. The decades up to the 1980s were particularly expansionary in that regard. Inside statistical outfits, macroeconomic statistics became the dominant power. Even such attitudes were moderated more recently. The rise and influence of *interest groups* and *sectoral approaches* lowered the focal point below the macro, although not quite to the micro. (Sometimes, this level is called *meso*, from the Greek for *middle*.) Microdata were still biding their time.

## The Power to Reveal

Records containing observations on individual units can be created for all sorts of things. They can document buildings, guns, or bacteria. Of course, socio-economic research focuses overwhelmingly on people and businesses. When microdata exist, they can support quantification at any desired level of aggregation, and turn data into information. Staring at all those census returns is pointless. Doing the work and announcing that the population is 30 million is something useful. Counting men and women by age, even more so. We get answers to our initial questions, but is it possible to go beyond that? Advanced planning for our research is the right thing to do, but our brains don't stop. Sometimes what *we want* is no more than what *we think we want*. There's nothing shameful about a thought triggered afterwards, as we saw in Chapter 7 with the example of age and income. The microdata already collected were sufficient, waiting for a question that hadn't been asked yet. It's a good day when that happens.

**FIGURE 8.1**

A look inside a population change between two periods

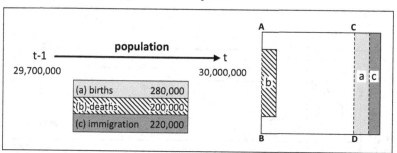

Microdata offer flexibility. At times, they can be used to reveal deeper layers of the truth we seek. Take a simple annual population increase. A release states that today (time *t*) the population is 300,000 higher than the same day last year (*t* – 1). This is newsworthy and important. But it doesn't reveal what *really* happened between these two points in time, a question that might come after. The overall increase might have come about by the events captured in the data in Figure 8.1. People in that country attended 200,000 funerals, witnessed 280,000 births, and welcomed 220,000 newcomers to the country. The population at time *t* – 1 (rectangle ABCD) is replaced by the bold outline at time *t*. Only the white part of the population box is common to both years, deaths (b) are no longer in the mix, and there's a half million new individuals from births (a) and new immigrants (c).* Whether we refer to such narratives as context, explanations, or granularity, they're crucial and require microdata. They also highlight the importance of using administrative microdata sources for births, deaths, and immigration. Even back-to-back annual censuses would not be superior. For instance, deaths between *t* – 1 and *t* would have to be inferred from the absence of those individuals in year *t*.

Microdata can also refine research findings when aggregate measures conceal *compositional shifts* that occur over time in the populations under study. Whether related to income percentiles, age groups, or a host of other aggregations, the implications for policy interventions are significant. Take poverty, for instance. Using income measures, cost-of-living estimates, residential locations and more, researchers establish *poverty lines* (or *low-income cut-offs*). Figure 8.2 shows two such poverty lines, 10 years apart (*t* – 10 and *t*), as the thick vertical lines in the population rectangles. It also shows that the number of individuals below the poverty lines remained the same between the two periods, at three million, but the *incidence of poverty* declined from 12% to 10%. This is so because the population increased from 25 to 30 million.

---

* In reality, it'd be a bit more complicated as *c* really refers to net migration.

**FIGURE 8.2**

The changing composition of the poor

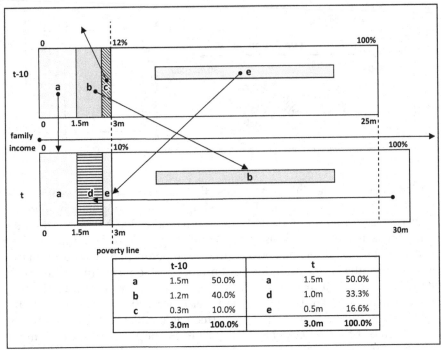

|  | t-10 |  |  | t |  |
|---|---|---|---|---|---|
| a | 1.5m | 50.0% | a | 1.5m | 50.0% |
| b | 1.2m | 40.0% | d | 1.0m | 33.3% |
| c | 0.3m | 10.0% | e | 0.5m | 16.6% |
|  | 3.0m | 100.0% |  | 3.0m | 100.0% |

If this came about because of well-publicized government interventions a few years ago, including specific transfers to families for each kid under 18, the headlines today would rightly point to some progress. But is that enough? Whose poverty is it? There's no poverty without the poor. Who are they?

Microdata make it possible to look at the composition of the population below the poverty lines in each of the two years. The parts in the shaded areas on the left of the population rectangles are different – see also the table at the bottom. From the three million poor in year $t-10$, half are still present in year $t$ (area a) experiencing chronic poverty. However, 1.2 million individuals (40%) lifted themselves out of poverty and in period $t$ they're dispersed among higher incomes (area b). The remaining 10% in $t-10$ (area c) don't exist anymore, attrition takes a toll too. In period $t$, one-third of the poor (one million) are new (area d). They came among the five million new individuals that accounted for the population increase, and they may be families with kids born in the last ten years or new immigrants (not shown – but all that can be found out too). Another half a million of today's poor (16.6%) were relatively well-to-do and in higher incomes ten years ago, but clearly they've fallen in hard times (area e).

This is only a teaser of the revealing capabilities of microdata. More detailed analyses can examine the relationship between individuals and families, follow the number and proportion of kids in poor families or the trajectories of new immigrants. For instance, a horizontal line in the shaded area in $t -$ 10 could show that 70% of the poor families have kids under the age of 18, whereas a similar line at time $t$ may show that this proportion has now fallen to 50%. Could this provide more credence to the child support program? Or answers to how best anti-poverty measures can address specific underlying causes, perhaps differentiating chronic poverty from more predictable short spells? Being fully aware that the make-up of poor populations is not the same over time matters. Otherwise, our understanding remains at a super-ficial level. As we've seen repeatedly, our times call for more accuracy. Well, microdata are the key to this.

When microdata were non-existent or out of reach, decades of research got habitually fixated on macro questions satisfied by summary answers, constructed any which way. For a summary statistic, no need to go far-ther than the *average*. Early theoretical advances in statistical distributions asserted that *statistical moments* fully describe the underlying data. In a moment of excitement, the mindset was even to keep the arithmetic mean and the variance and throw out the microdata behind them. Thankfully, this kind of *information management* never caught. In our example of income by age in Chapter 7, the microdata produced an average income for that particular group of $48,354. Not satisfied with that summary 'truth', or to answer the new question by comparing the incomes of the young and the old, the microdata were rearranged in several pieces (age bands). Arithmetic means and variances can be computed independently for each piece. Some arithmetic means will be higher and some lower than the overall average, all individual variances will be lower than the group's total. We can keep at it, drilling down to more granular levels, by making the bands smaller. The information-to-data ratio will continue to rise. But why not take it all the way? We can plot the complete distribution of microdata. The totality of the information is there. Not using it is jewellery forgotten in a safety deposit box.

Here's how the microdata really look like (Figure 8.3, with the relevant part of Figure 7.5 reproduced as an insert for convenience). They contain the whole truth collected for that group, each person's age and income. Names are not necessary. Visual inspection can discern that generally, in this par-ticular group, incomes increase with age up to the age of 50 and then fall – but not to levels as low as the incomes of those under 20. Several individuals in their late 20s and 30s earn more than individuals in their 60s and 70s, yet a couple of people in their 70s still have higher incomes than many in all younger groups. Surely, this scatterplot is not as clean and neat as the insert. It doesn't have straight lines and orderly columns nor does it include stylized age bands with averages – all of which are made up by line drawing. It's

**FIGURE 8.3**
Microdata vs. aggregates

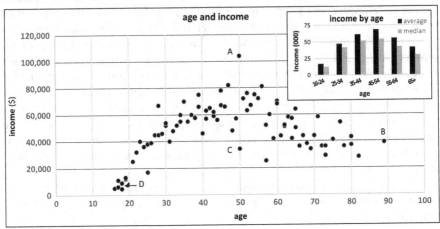

messier, with dots scattered all over. But that's how the plain truth looks like! Those dots are not only capable of answering the question posed but they offer so much more.

They contain a lot of extra information to answer questions that haven't been asked yet. They'll uncover that the highest income earner is a 50 year old (A), reveal his exact income ($104,000), and show that it's the only one in excess of $100,000. They'll show that another individual of the same age (C) makes a third of that ($34,200), if proof-was needed that there's more to generating income than someone's age. They'll reveal the minimum income (D, $4,600) and that few individuals' incomes are identical. They'll also 'identify' the oldest individual in this group (B), at 89. The insert hides so much truth and provides 'lesser' answers to our questions. Why settle for less in our demanding times?

Microdata from any source can be exploited in such analyses. Complete, census-like microdata sets in particular can illuminate with confidence the darkest corners of the research space and make us assertive in communicating findings like those above. We can even create trivia games too. We can ask "*Who's the wealthiest person in the country*", "*What's the age of the oldest person alive*", "*How many centenarians live among us today*", "*How many babies were born on New Year's Eve*", "*What's the biggest number of kids in a family*" or tell the world that almost 70% of Canadians really live within 100 kms from the US border and not in the North Pole. Practically, imaginative uses of microdata are constrained by confidentiality. Disclosing the identities of the richest and the oldest is a no-no, even though chances are that such announcements would have already been all over the place.

In any event, we're moving steadily towards higher-information environments for a while now. Blanket generalizations no longer suffice and more granular data set the tone in research, policymaking, and many business practices. A familiar example comes from insurance companies and the use of actuarial tables for life insurance or the establishment of premiums for property insurance. As microdata became more abundant over time, it became possible to construct probabilities at more granular levels. Pricing schemes supported by such data create winners and losers. For instance, 20–25-year-old males are more prone to car accidents and pay higher premiums. If a Figure like 8.3 had coordinates that showed age and accidents, a 23-year-old who can't afford the going rate could pinpoint his own dot and argue that he's only come close to a fender-bender once – and then he wasn't at fault. He'd point to D in the chart and ask for leniency – before he goes nowhere and settles for a car sensor to monitor his driving. But you can see why he may advocate for individual treatment and Figure 8.3, while someone else his age would gladly take cover behind the insert. If we believe that the truth will set us free, example after example will show that the path is 'microdata all the way'. The danger from now on is the embarrassment of riches. With so much accumulating, our scatterplot could look like an amorphous black blob and feel like gobbledygook.

Use of microdata demonstrates that there's loss of information as we move up the aggregation pyramid. Not only can aggregation hide useful signals but at times it can even lead to conclusions entirely reversed by more detailed constituent data, if they were available. Consider the numerical example of Table 8.1. If X stood for men and Y for women applicants for positions in the military or a university, and the columns indicate acceptance against an equal number of applications, the data clearly show that a smaller proportion of women were successful compared to men (70% vs. 75%). However, if the three groups that make up the total were distinct branches of the military (air force, navy, army) or university departments (engineering, political science, psychology), a closer look at the data would reveal that women were proportionally more successful than men in each and every process (compare totals for each group). This phenomenon, known as *Simpson's paradox*, is a

**TABLE 8.1**

Reversal of findings or trends at different levels of data aggregation

| Group | X | | | Y | | |
|---|---|---|---|---|---|---|
| | Successes | Attempts | % | Successes | Attempts | % |
| Group 1 | 42 | 50 | 84.0 | 9 | 10 | 90.0 |
| Group 2 | 20 | 30 | 66.7 | 28 | 40 | 70.0 |
| Group 3 | 13 | 20 | 65.0 | 33 | 50 | 66.0 |
| Total | 75 | 100 | 75.0 | 70 | 100 | 70.0 |

manifestation of the disappearance or outright reversal of findings or trends at different levels of data aggregation. How's that possible? Is it indicative of some pliable rather than immutable truth? No wonder that the existence of such cases troubled truth-seekers, as well as provided fodder to cynical views of statistics, particularly concerning the hazards of causal inferences based on probability. Yet, it's well known that associations between variables can be affected by other unobserved, uncontrolled, or confounding variables. Other times simple arithmetical peculiarities can be at play, such as the concentration of high success or failure rates in small population subsets. Even before the deployment of statistical inference techniques to deal with this matter, it highlights the significance of knowing the data generation process and the materiality of aggregation. Moreover, it points to the need to go beyond any immediate data set at hand. In an expanded context, pertinent questions would include why only 10% women applied in the air force or engineering (Group 1) compared to 50% in the army or psychology (Group 3).

The context will also explain when and why more detailed data may muddle rather than clarify the truth. To see this by way of the same Table 8.1, make X and Y two individuals who compete on a test with 100 questions. They do so over three days, at their own pace. This time the rows refer to the correct answers in the number of questions each individual chose to answer each day. The scores indicate that individual Y beat X every single day (totals for each row). However, clearly X outscored Y with a 75% vs. a 70% average. Oddities like this can arise because of the different rates of success in different-sized subsets of questions. If questions were answered in order, it's clear that in the first two days Y answered 37 out of the top 50 (74%), which is lower than the 42 questions (84%) of the same set answered correctly by individual X during the first day. Both individuals answered correctly 33 questions out of the bottom 50, for an average score of 66% (which is the weighted average of the last two days that individual X took to answer them). There's no ambiguity as to the end result, the overall score is what matters and it speaks for itself. It may be tempting to discard all microdata as irrelevant and unwanted noise. However, even here, they may still be of value. If the test had sections covering different topics, such data could reveal the comparative performance of the competitors in those parts. In the limit, this can go all the way, one question at a time.

The fact that microdata can be scaled up to support higher-level aggregates points to the desirability of a bottom-up approach. Top-down macro indicators will always pose limitations. Moreover, when incomplete microdata only exist to prop up the production of some aggregate, drilling down can only go so far. Even less mileage can be had when trying to combine microdata from accidental overlaps, such as among different survey samples. It's precisely such matters that gave rise to longitudinal (panel) surveys that follow the same population over time – with all the problems of attrition and the like. Microdata based on complete registers could accomplish so much more, so much better. So, efforts in a direction that enable bottom-up statistical constructions need to be encouraged and supported. Initiatives that go against should be scrutinized, and any green light should be accompanied by conditions for improved long-term planning.

## Coming of Age

For the longest time microdata lived in the twilight of statistical activities. They were assigned supporting roles for macro aggregates, frequently ignored, neglected, or looked down on, and occasionally they were outright abused. The absence of microdata from the research scene curtailed the possibilities space, acting as a brake on imaginative ideas. At times, researchers wouldn't even dare think of designs based on microdata since they weren't available. When they were available, they weren't accessible. In the few cases that they were available and accessible, they weren't clean! Scrubbing microdata is not a glamorous part of research. But this is no longer. Microdata would not be denied their rightful place. Such power can't be held under wraps for long. The vision of ever-expanding statistical registers alone is a testament to the rise of microdata.

### Statistical Registers

What truth can be learned from microdata is limited only by the contents of their individual records. This points to an important distinction between census-like and incomplete microdata sets. There's only so much that can be done with microdata from sample surveys, for instance. The new research paradigm works best with complete, curated, and well-maintained microdata. Then, research is not held hostage to the confines of some initial design, regardless of how well thought-out it might have been.

The epitome of such data is the *statistical register*. A complete and exhaustive collection of all units in a population. Every individual in the case of a population register, and every business in the case of a business register. The construction of registers requires photographic-type stocktaking, followed by reliable and ongoing feeder sources both for additions and subtractions. Their quality depends crucially on timely updates, whether in real time, sub-annual, or annual intervals – but certainly not waiting for a decennial census. Depending on the type of register, vital statistics, tax records, postal addresses, bankruptcy files, building and demolition permits, all would be examples of such feeder sources.

The stereotypical portrait of a register is something like an old telephone book or a directory. The notion used here is conceptually much wider.

Tombstone information, such as name, address, and date of birth, is only the tip of the iceberg. Registers are conceptualized as data-thirsty, bottomless pits that can absorb any and all data related to their subject matter. They're more like linked data environments, capable of expanding outwards with the addition of more sources. In the spirit of our previous free-willing discussion regarding no boundaries between data sources and linkages, each additional source becomes a new 'field'. For instance, there's no reason why a business register cannot include financial and employment data, identify exporters, R&D performers, or recipients of business support programs. Similarly, a population register can incorporate all those with a passport, and identify married, divorced, and widowed individuals. Even survey data can be linked, albeit they won't be as powerful since they don't represent complete microdata sets.

With such register environments up and running, our conceptualization can be extended to include coherent linkages among them leading to data integration. Individuals can be matched to households. Working individuals can be paired with their employers in the business register, which in turn must contain trustworthy hierarchical relationships between enterprises, their establishments, and their individual locations.[†] Both households and businesses can be linked to a building register. Then, not only we'd know the activities of businesses but we'd be able to place them on a map. A particular business may have a manufacturing plant, a warehouse, a head office, and a retail outlet. Another one may occupy an office in a nondescript floor of a high-rise. Such integrated environments will be the equivalent to zooming-in at high magnification in the network of nodes and edges discussed earlier. The multitude of relationships and interactions among nodes would come alive.

The time dimension can be dealt with easier in some data environments than others. As we've already discussed, business births and deaths are not as straightforward as those of individuals. Buildings have their own peculiarities due to non-unique addresses, extensions, or alterations. With good microdata available, and appropriate decisions, time series fit for most uses can be constructed. Researchers can even craft alternatives if the intended use warrants the effort. Lastly, under any information management framework, pertinent histories must be preserved in the form of older versions. Census data are used in genealogy and other studies when made public, decades later. Specific research needs could rely on past data as well of the blockchain variety. Studying the influence of R&D in wireless technologies won't benefit from knowing that NOKIA started as a pulp mill more than a century ago, and was active in the rubber business, but the company's R&D spending at the time when it left its mark as *the* major player in his area would be critical.

---

[†]   Seen as information sources, up-to-date registers will be crucial in the future. At present, business registers are nowhere near real time. When openings and closures were needed under COVID-19, Google locations were used. Relying heavily on information supplied by individual owners, they weren't very reliable.

This all may sound daunting but there's no overriding reason why such foundational statistical infrastructure cannot be the vision driving upcoming efforts. Some countries are well on their way. More importantly, we now have a map and know where we stand on it. Therefore, we can plot a path for any destination we choose.

## Clinical Trials in Economics

Coupled with more ambitious demands and improved technical capabilities, at last the stars are aligning for microdata. Many, new, and important research possibilities are opening up. When called upon, microdata deliver. Watch this space – with an example.

Governments dole out significant sums of money in support of businesses. The dispersion of funds is administered by different departments through programs aiming at a variety of outcomes, such as employment creation, technology adoption, productivity gains, or clean tech investments. Support assumes many forms, including grants, loans, equity financing, or other repayable or non-repayable contributions. An overarching research question has always been what good comes out of all this spending. Are the stated objectives met, and to what extent? Perhaps employment in recipient firms goes up for a year or two but what happens five years out? Does the support help the survival of weak firms or delay the inevitable? At the end of the day, should such programs be maintained, expanded, or are there better uses for the money? In a daring day, someone may even raise the issue of which program performs better.

Program administrators habitually performed the occasional evaluation of their individual program, whether because they had to or they wanted to. They possessed detailed records of the support they gave to individual businesses. But they had no information on what mattered most, the performance of the recipient businesses *after* the support. Historically, such evaluations relied on some 'key informant' interviews or sample surveys among a few recipients. What else could be done without an independent and reliable source of information?

Fast forward to our days to see how everything we've discussed is coming together. First, there have been noticeable quality improvements in business registers, largely due to authoritative administrative sources with complete population coverage. At a minimum, they can produce business demographic statistics. How many businesses are active at a given time, in which industry, and what location? A longitudinal business register can go beyond isolated snapshots and answer how many businesses were created since last year, in what sector, and where. (This is still work in progress even in advanced countries due to difficulties with business time travel.) The new data sources came with reinforcements that enabled business registers to expand, as we described. Tombstone information was complemented with financial, employment, and other data. Key variables that can track firm performance, such as revenues, expenses, profits, and employment, have now become readily available. Additional data sources continued to be developed and added, for example, exporter flags. You can see the transformation. From struggling survey frames to powerful information sources in their own right.

This whetted appetites and further expanded the possibilities space. The surveys undertaken as part of the evaluations mentioned above went after sub-populations of recipient businesses asking for the type of information that now resides within modern business registers. Such surveys weren't mandatory and never worked well, but why bother anyway? This information is now available, and for *all* businesses. Therefore, in the spirit of the new research paradigm, the next step came naturally. Organize the individual programs' own data and use linkage keys to identify recipient firms in the business register. Then, the needed performance indicators can be extracted for each firm in the set. Revenue growth, employment, and profitability have always been indicators of interest in such studies but even more is available to be had, such as export performance or market shares. Productivity indicators can be constructed too, from simplistic output-to-labour ratios all the way to multifactor productivity. Practically, the exercise to integrate data from business support programs to the business register involves both *deterministic* and *probabilistic linkages*, particularly if the programs don't have business identifiers other than names and addresses. Match rates in such exercises are commonly used as data quality indicators, and a 90% is considered high. Recalling previous comments on unique identifiers, there's no reason why not to raise the bar in the future to 100%. For that, statistical evolution depends on such program areas too. The new paradigm works best if cultures change and all players aim at the same target. Program data collection can incorporate from the get-go information that would facilitate subsequent linkages.

Once linked, program data too become integral parts of an expanded business register. They're census-like, in that they represent all businesses supported by a particular program. Sure enough, in a few countries such research has started. Not only does it take care of comparative performance assessment and program evaluation functions, but it does so in an unmatched way. From the realm of dreams earlier, it's now an intuitive and seemingly simple approach. No need for expensive and time-consuming surveys, no response burden on businesses, better quality data, and an expanded business register to support future uses. Win-win all around.

This real-world example sums up our discussions well. But there's quite a bit more to this story. Thus far, we arrived at a point where we can easily match the evaluation studies as they were done up to now. Finding out that recipient businesses had 5% annual revenue growth and 4% employment growth for three consecutive years after a grant may shed some useful light in the research. But, at best, such performance can be compared with the performance of the same businesses before the grant or some industry average. Now, we can do so much better. A century later, we can follow what Fisher preached in the 1920s and design randomized controlled experiments. What about comparing the performance of the recipient firms with that of similar firms who didn't get the support? This is considered the holy grail in medical research and clinical trials for decades but for socio-economic phenomena even the thought wasn't contemplated. The absence of microdata was limiting possibilities.[‡] We may have a treatment group but where do we find a control group that received dissolved sugar and measure the placebo effect?

This is slightly tricky, enough to warrant a quick explanation. To settle matters conclusively, the ideal comparison would be between how well someone (a patient, a business) does with an intervention (pill, grant) vs. how well that same someone would have done without it. This is impossible to ever observe. It's a purely hypothetical trajectory that can never be known (except in movies with parallel universes), an unknowable unknown. The best we can do is to compare what happens to someone subject to the intervention with someone identical without the intervention. Rather than *identical*, practically we settle for *as similar as possible*. The implicit assumption is that since the performance of both was very similar prior to, and up to the point of, the intervention, any subsequent difference can be reasonably attributed to the intervention. In clinical trials, genetically identical twins would be ideal, but we can't have that in businesses.

Let's see how this microdata research works in practice, and what steps are involved. First, the list of firms in a program are matched with firms in the business register and the relevant program data are appended as new variables ('linkage'). They're now part of the business register and will stay forever. These businesses form the *treatment group*, that is, they were recipients of the grant (used generically for any support). Second, data for all desired variables are selected, and for all businesses in that program. Depending on the scope of the research this step can be repeated for each program, undertaken for the totality of programs, or for any chosen grouping of programs. Third, the control or *counterfactual group* is constructed, preferably of equal size. This is the most sensitive step and will make or break the findings. The idea is to select firms as similar as possible to those in the treatment group so that going forward any differences in performance can be attributed to the grant, not other covariates. Dimensions typically used to determine similarity include industry, firm size (revenues, employment), and performance (profitability, productivity). Additional criteria can also be used as warranted, for example, exporter status or R&D performer. Practically,

---

[‡] Studies using this methodology became the standard in medical studies and clinical trials, where researchers generated their own microdata. At times, they also come in the double-blind variety where neither the researcher nor the patient know who's on the drug and who's on the placebo.

researchers have methodological choices, such as closest neighbours and propensity scores, and sophisticated software tools to arrive at a control group. With the two groups identified and the data sets complete, research proceeds as usual. Descriptive analysis is followed by inferential analysis employing econometric techniques such as differences-in-differences, regressions, and the like. Allowances are made for lagged effects, as it may take time for causal influences to materialize or peak.

Such exercises have been made possible because of the existence and increasing maturity of microdata. The fact that they have begun in a handful of countries is a sure sign that microdata are coming of age. Once their potential is unleashed, there's no stopping it. New uses will be found to enable more.

## Playing at the Margins

In the example just discussed, the firms in the treatment and the control groups were similar in most respects. They were separated only by the grant. Presumably, the program criteria would have been applicable to all. Why then did firms in the control group not get the grant? Either they didn't apply or they didn't succeed. The latter case holds particular promise for more detailed research. Typically, grant programs approve or reject applications on the basis of a scoring threshold. Except for those clearly over or under that threshold, there are always applicants who succeed or fail marginally. Comparing these otherwise similar groups can shed even more light on the impact of the grant on the comparative performance of recipient and non-recipient firms. Even thinking of such possibilities has implications for data. Some programs keep the applications and the scoring data, others discard them. The feedback loop from the research to the programs suggests that, if such studies are to be done, program procedures must be aligned with the needs of microdata research. Collecting business numbers as identifiers for the linkage exercise and microdata on near misses would facilitate the creation of such data sets.

This links back to our discussion in Chapter 5 concerning the drawing of lines. Near successes and misses will ensue as an inevitable consequence of the decision to draw a line. Someone or something will be almost touching the threshold, but from either side. There are countless real-life situations where marginal differences between pass or fail can be so critical as to determine someone's future path in life. In the education system, differences between a B and a B+ and cut-off scores in standardized tests determine admissions to college and future careers. In health care, cholesterol results just above or below a certain measure lead to different treatment options. Banks use similar lines to determine eligibility for a loan or a mortgage. It's always been a challenge to understand the longer-term implications of decisions made from such seemingly arbitrary metrics.

This brings to the fore the issue of initial conditions. Chaos theory made a splash in the 1980s by focusing people's imagination on endless, *fractal*, loops. A simple equation like $x_{t+1} = kx_t (1 - x_t)$ can produce weird things. Other than the constant $k$, the only term is $x$, whose next value $(x_{t+1})$ depends on its current value $(x_t)$. So does the one after that, and so on. Depending on the value of $k$, though, a population represented by $x$ can achieve a steady state, bifurcate indefinitely, or contract to oblivion. As we've seen, there's no such thing as a straight line – and our simple equation is, of course, non-linear. The essence of the theory is that chaotic behaviour can be brought about by minuscule differences in initial conditions (the *butterfly effect*). Starting points, even next to each other, matter enormously. They can lead to totally different outcomes over time. This is not a trivial and esoteric matter. While a fact mathematically, we need to question to what extent, where, and how it applies in real life. Historically, much of our current knowledge that leads to corrective or preventive actions is based on findings from studies that focused on outcomes associated with groups of 'ins' and 'outs'. The population straddling both sides has been ignored due to the lack of microdata. It'd be quite revealing to compare across marginal differences in initial conditions, those close neighbours that touch the line from opposite sides. Clearly, such research requires detailed microdata.

There's been enormous policy interest in the returns to education. Over the years, studies have shown that the education premium is quite high – it pays to go to school. In the US there's also been a lot of research on premiums associated with the quality of the university or college attended in conjunction with the massive student debts accumulated. To answer such questions, microdata looking at admission cut-offs that rely on high-school grade point average (GPAs) and standardized scores, notably the scholastic aptitude test (SAT), can be quite useful. A strand of studies with a methodology known as *regression discontinuity design* have demonstrated that access to such microdata is possible.

A 2009 study by Mark Hoekstra[1] addressed the question of whether admittance to a state flagship university can boost future earnings by comparing students who were barely admitted against applicants who were barely rejected based on the admissions cut-off. The (undisclosed) university retrieved data for every white, male applicant over a three-year period, 1986–1989, with personal information, test scores, and whether he was actually enrolled. The data set included high-school GPA scores, adjusted for different scales used in high schools. The university then sent these data to a state office where, using social security numbers, they were matched to records with earnings from 1998 through 2005, that is, 10–15 years after high school. The study found that the earnings of those enrolled at a flagship state university were approximately 20% higher relative to applicants who were barely rejected. However, the data set contained no information on whether rejected applicants ultimately attended another university.

In 2020, Smith, Goodman, and Hurwitz[2] produced estimates of the economic impacts associated with enrolment in all four-year public sector universities in Georgia, attended by about half of the state's high school graduates. The SAT thresholds used for admissions by all colleges allowed comparisons of otherwise identical students who differed only in their college options. If the state's four-year colleges were not an option, most students below the cut-off for enrolment in the state's two most selective flagship universities would not have enrolled in a four-year institution at all. Some would have enrolled in a two-year college or no college at all. The researchers matched SAT data from the College Board for high school student cohorts graduating between 2004 and 2008 with data from the National Student Clearinghouse, which tracks enrolment and completion in post-secondary institutions across the US. Then, they linked the data to the TransUnion credit bureau data to obtain a snapshot of the students' household incomes in 2017, which includes the student's own wage earnings, spousal earnings, and non-wage earnings. The study found that enrolment in public four-year colleges boosted household income around age 30 by 20%, driven almost entirely by students from low-income high schools. It also found that such enrolment had little impact on student loan balances as those who barely met the admissions criteria and those who barely missed them had similar balances at age 30.

In 2014, Seth Zimmerman studied the returns of college admissions for academically marginal students in Florida.[3] Students with high school grades just above the admissions eligibility threshold at a large public university were much more likely to attend any university than students below the threshold. Through an agreement with the state's Department of Education, he obtained data for six cohorts of public school 12th graders from 15 counties, who graduated between 1996 and 2002. These administrative data included basic demographics, high school, community college, and state university transcripts, as well as application data for all 11 state universities. These data were linked to SAT records provided by the College Board and to state earning records from 1995 to 2010. The study found that, on average, students induced to attend college by 'threshold-crossing' attended a state university campus for an additional 3.8 years, and they graduated at rates similar to those in the broader student population. The marginal admission yielded gains in earnings of 22% between 8 and 14 years after graduation from high school. These gains outstripped the costs of college attendance and were largest for male students.

In a 2016 study of the same strand, rather than the returns to marginal college admission, Ben Ost et al. examined the returns to *college persistence* among admitted students.[4] They looked at low-performing Ohio students on the verge of being dismissed from four-year universities due to poor GPAs. Students are put on probation if their cumulative GPA falls below some critical threshold (typically below 2.0), and dismissed if they fail to raise it in subsequent terms. Using enrolment data from 13 state universities, state earnings, and GPAs, the researchers compared mid-run earnings of students with GPAs just below the dismissal cut-off to those just above. They found that low-performing students who were dismissed suffered substantial earnings losses, measured between 7 and 12 years after college enrolment. Students just below the cut-off were approximately 16 percentage points less likely to enrol in the following term, completed 0.2 fewer years of school, and were 10 percentage points less likely to obtain a degree relative to those above the threshold. This correlated with a decrease in earnings of approximately 4.5% between those just below the GPA cut-off compared to those just above. Persisting and graduating likely provides a large positive return compared to dropping out, even for low-performing students.

In the book *Everybody Lies*, Seth Stephens-Davidowitz describes the case of Stuyvesant, a prestigious public high school in New York with the highest SAT average in the state, and which requires the highest score for admission in the exam-school system.[5] Competition is fierce since entrance is seen as a ticket to success in life, leading to higher SAT scores, admission to Ivy League Universities, and better careers. But is that so? Allegedly, too many people will always remember near misses and near successes, by a point or two. The author correctly remarks that comparing those who went to that high school with those who didn't is not where we should look for the answer as, by definition, those who made it are an elite group of high-achievers. He recommends a regression discontinuity study. Comparing marginal differences in scores around the cut-off would look like a randomized controlled experiment, with those just above being the treatment group and those just below the control group.

A study to evaluate such causal effects was actually carried out by Abdulkadiroglu et al. in 2014.[6] It exploited admission cut-offs in the three original exam schools in New York spanning grades 9–12, and in Boston's three exam schools, spanning grades 7–12. The study combined several years of data from applications, test scores, registrations, and attendance records with college enrolment from the National Student Clearinghouse. It found no peer effect in either system, that is, there's little evidence that exam schools boost achievement or they have any effect on eventual test scores and admittance to better quality colleges. To be fair, unsuccessful candidates probably went to pretty good schools too. The study did find some evidence, though, that successful students can expect to study with fewer non-white classmates, and that house prices jumped in those school districts!

In addition to the comparative study of populations adjacent to cut-off lines, similar microdata can help the very establishment of such cut-offs too. For instance, creating an age group of 20–25 year olds for car insurance purposes could benefit from sensitivity testing involving those aged 19 or 26. Microdata can make significant contributions in research around the margins. It could provide a challenge function to results that may have led to misconceptions or the wrong stereotypes.

## Deep and Dynamic

As microdata mature and time series accumulate they'll enable research that can improve the quality of findings in areas where the composition of the populations matters. These extend to intergenerational studies. Until now, such work relied on aggregate data, longitudinal surveys with small samples, or the painstaking and patchy construction of limited data sets from administrative microdata struggling with timing, cohort, and other hurdles. However, even those proved revealing.

For example, a 2000 study examined matched parent–children recipients of unemployment insurance in Canada and Sweden through panel data sets. In Sweden, the data set was built from a sample of the population register. In Canada, individual tax returns were linked to families and then children were linked to parents by name and address through a series of steps.[7] These data sets were then linked to unemployment insurance claims generating observations for just over 6,000 individuals in Canada and almost 4,000 in Sweden. The analysis showed that the incidence of using unemployment insurance was high among young adult men in both countries, but it was the linkage to parents that proved more revealing. In both countries, young men whose fathers had claims in the past were more likely to also have claims, 80% vs. 70% in Canada and 70% vs. 58% in Sweden. Moreover, in Canada, individuals whose fathers had past claims generally began their first claim sooner, while in Sweden the probability of repeated claims among individuals who relied on the program increased. Intergenerational learning at work, passing the torch.

Another example of dynamic analysis, this time based on business microdata, comes from a recent Statistics Canada study.[8] SMEs are considered the engine of the economy and governments are interested in their sustainability and growth. Discussions involving business size, though, commonly take place in a context devoid of the changing composition of such size categories, much like we saw in discussions concerning poverty. Policies relevant to businesses with less than 50 employees, for example, are implemented based on the number of businesses in that category and their proportion of total employment. However, identifiable microdata make it possible to have an instructive look at the changing composition and the growth trajectories of individual businesses. All active statistical enterprises with at least one employee (more than one million) were decomposed to many employment size classes over two different times, 2008 and 2014. This allowed the production of Table 8.2, shedding light on interesting transitions.

The diagonal shows that roughly half the enterprises remained in the same size class over this six-year period. The precise figures ranged from a low of 38.9% for enterprises with 5–10 employees to two-thirds for the largest enterprises with more than 1,000 employees. This means that more than 60% of the enterprises with 5–10 employees in 2014 were not there six years earlier! Some enterprises moved to a higher employment class during

**TABLE 8.2**

Transitions of enterprises by size class over time

Enterprise Size Class Transition Matrix for the Period 2008 to 2014

| Enterprise Size Class, 2008 | Enterprise Size Class, 2014 | | | | | | | | | |
| --- | --- | --- | --- | --- | --- | --- | --- | --- | --- | --- |
| | Between 1 and 4 | Between 5 and 9 | Between 10 and 19 | Between 20 and 49 | Between 50 and 99 | Between 100 and 249 | Between 250 and 499 | Between 500 and 999 | 1,000 and above | Attrition[1] |
| | % | | | | | | | | | |
| Between 1 and 4 | 50.0 | 5.3 | 0.8 | 0.2 | 0.0 | 0.0 | x | x | x | 43.7 |
| Between 5 and 9 | 19.9 | 38.9 | 10.5 | 1.2 | 0.1 | 0.0 | x | x | x | 29.2 |
| Between 10 and 19 | 5.1 | 15.5 | 43.5 | 12.1 | 0.6 | 0.1 | 0 | x | x | 23.0 |
| Between 20 and 49 | 2.3 | 2.6 | 13.3 | 51.5 | 8.6 | 0.8 | 0.1 | 0.0 | 0.0 | 20.7 |
| Between 50 and 99 | 1.6 | 0.9 | 1.7 | 16.8 | 47.7 | 11.4 | 0.5 | 0.1 | 0.0 | 19.3 |
| Between 100 and 249 | 1.0 | 0.8 | 0.9 | 2.6 | 13.4 | 52.7 | 7.6 | 0.7 | 0.1 | 20.2 |
| Between 250 and 499 | 0.6 | x | x | 1.4 | 1.7 | 13.8 | 45.9 | 11.1 | 0.9 | 24.0 |
| Between 500 and 999 | x | x | x | x | x | x | 10.7 | 48.7 | 12.1 | 22.4 |
| 1,000 and above | x | x | x | x | x | x | 1.4 | 4.6 | 65.9 | 25.9 |
| Addition[2] | 74.9 | 14.4 | 6.3 | 3.2 | 0.8 | 0.3 | 0.1 | 0.0 | 0.0 | ... |

*Source:* Entrepreneurship Indicator Database, Statistics Canada, Centre for Special Business Projects, Table 3, www150.statcan.gc.ca/n1/pub/18-001-x/18-001-x2018001-eng.htm

x: Suppressed to meet the confidentiality requirements of the Statistics Act.

...: Not applicable.

1: An enterprise belongs to the size class 'Attrition' when it is in one or more of the following situations: has no employee, has ceased operations, or has merged.

2: An enterprise belongs to the size class 'Addition' if it did not exist in 2008 as an employer, didn't exist in 2008, but existed in 2014.

that period, whereas others transitioned to lower employment. Many more were subject to 'attrition', as shown in the last column. Clearly, this is significant. Among the smallest enterprises in particular, with 1–4 employees, 43.7% of the 2008 cohort were no longer around – such is the fate of many small businesses under the forces of 'creative destruction'. New enterprises are also created continuously, and the majority start very small. The last row shows that enterprises with 1–4 employees accounted for three-quarters of all additions between 2008 and 2014. Since all the data exist, such analysis can be extended to include further breakdowns by individual industry and other characteristics, painting a really detailed profile of the changing composition of businesses. For instance, attrition was lower in manufacturing than in the service sector. Understanding the evolving structure of an economy under dynamic conditions adds a valuable perspective compared to general platitudes. Granularity in the data can improve policy by making it more granular too.

## Research Problems and Data Stewardship

Our interest in this chapter has been on microdata and the research possibilities they open up. Research methods and findings themselves are different matters altogether. Poking at such matters, though, would not have been possible without microdata in the first place. The only alternative would have been anecdotal accounts, always prone to doubt.

Problems surrounding research results are well known and, of late, openly discussed. There have been cases of fraudulent or made-up results, some high profile, many more lesser known. Such misconduct aside, other well-known matters cast a shadow on published research too. From journal preferences for positive results, which bias the overall balance of findings since failures are barely reported, to the methodological overuse and misuse of significance tests (p-tests in particular), to the hogging of microdata crucial to replicate studies. Such matters are particularly true when research generates its own microdata, in experimental settings and clinical trials.

A good deal of attention is now dedicated to *meta-analysis*, systematic reviews of multiple previous studies. Focusing on medical research, such a paper by John Ioannidis in 2005 gave quite a shake to the system by arguing that most research findings are false.[9] He asserted that "...*a research finding is less likely to be true when the studies conducted in a field are smaller; when effect sizes are smaller; when there is a greater number and lesser preselection of tested relationships; where there is greater flexibility in designs, definitions, outcomes, and analytical modes; when there is greater financial and other interest and prejudice; and when more teams are involved in a scientific field in chase of statistical significance*".

The preoccupation with *p*-values as the appropriate statistic to accept or reject a *null hypothesis* has been severely criticized. Null hypotheses ($H_o$) effectively assess the plausibility that outcomes come from some intervention rather than chance alone. Quite simply, $p$-values are the probabilities used to make such decisions at a certain level of confidence (most times at $p < 0.05$). If there's less than a 5% probability that the difference between treatment and no treatment is zero we don't reject the null hypothesis – we have 'statistically significant' results. However, a null hypothesis can only be rejected or not, never conclusively accepted – if rejected it opens the door for the alternative hypothesis ($H_1$). This will be discussed more in Chapter 9 but, for our purposes here, the main point is that 'statistically significant' results based on $p$-values are favoured for publication. This lead Carl Bergstrom and Jevin West in *Calling Bullshit* to also refer to *p-hacking*, effectively manipulating data or processes to get the 'right' $p$-value. As they put it, though, to measure the publication bias in favour of positive results, the relevant question is what fraction of negative results get published. Alternatively, how much research in a certain area is not published and why? Thankfully, there is a way to look at that – in medical research. It involves, again, microdata.

Studies involving clinical trials in the US must be registered with the Federal Drug Administration (FDA) from their inception to their completion, regardless of publication. These microdata can be combined with publication data and study results to facilitate meta-analyses. In one such study, Turner et al.[10] looked at research on antidepressant agents involving 74 registered studies with 12,500 patients. From the 51 studies that were published, almost all (48) contained positive results, something that could paint a rosy picture for the medication. However, their analysis found that only 38 of the registered studies contained positive results (about half), of which 37 were published. The remaining 36 trials had negative (24) or questionable results (12). Only three of the studies with negative results were published, while 22 studies were not published. What's more distressing, 11 studies with negative or questionable results were published with a positive spin – the conclusions in the abstract were not consistent with the end results.

Another meta-analysis study by Roest et al. examined reporting biases in double-blind, placebo-controlled trials on the pharmacologic treatment of anxiety disorders. The study found evidence for publication bias, outcome reporting bias, and spin. From the 45 published articles, 43 (96%) showed positive results. This inflated drug efficacy since FDA data revealed that from a total of 57 trials, 41 (72%) had positive results. Of the 16 studies with non-positive results, three were published in agreement with the results reported to the FDA but another three were published with a conflicting conclusion. The paper states that *"Trials that the FDA determined as positive were 5 times more likely to be published in agreement with that determination compared with trials determined as not positive"*.[11]

Assessing if cherry-picking new or positive findings can possibly be a 'rational' response rather than a bias, complete microdata that include all research efforts in a given domain and allow for such meta-analyses offer some consolation. However, hardly anything comparable exists in socioeconomic research.

An additional salient point from this discussion is that research has a way to find entrance points to the microdata so long as they exist, notwithstanding their sensitivity. Diligence and measures to protect confidentiality and privacy will always apply to microdata. Such matters necessitate further deliberations around data stewardship and conditions of access rather than ownership. Perceiving data as a strategic resource will inevitably force fresh thinking on the role of custodians and the evolution of best practices related to information management, including data retention. While such issues were in the rear-view mirror thanks to the emphasis on the macro, they are now in plain view. A recent example comes from the annual report of the Canadian Statistics Advisory Council at the end of 2021, which calls for an updated Statistics Act fit for a *"modern digital society"* and which will *"clarify and strengthen...Statistics Canada's data stewardship role"* and *"introduce a new category of accredited users from government, academic and private research institutions, and Indigenous organizations and communities who would be granted access to more disaggregated microdata without having to be deemed employees of Statistics Canada"*.[12]

To conclude our discussion, there's a pecking order in the kingdom of data. In that realm, microdata are the king, and registers are his regents. Macro indicators are the kingdom's ambassadors that showcase its prowess everywhere.

---

## Fact-Checking Tips

*Spot the difference:* Finding differences between two seemingly identical pictures or cartoons has been a popular pastime activity. At first glance, the man sitting on an armchair reading a book, the dog next to him, and the painting on the wall check out well. Nothing jumps out. It takes careful examination to spot that the armchair has three legs, the man's left shoe has no lace, the face in the painting is missing an eye, and the painting is not hanging from a nail.

The macroeconomist's *big picture* is constructed from a large collection of data points. Defects on individual data points are neither immediately visible nor frequently enough to obscure the image. You'll come across many aggregates whose exact construction is unknown, but for which you have a reasonable idea of the microdata that make them up. Industry statistics will be based on individual firms, the CPI on the price of the items it contains. Movements in such aggregates may pass unnoticed if they roughly correspond to movements in their microdata, you'd feel better if the signals are at least in the same direction. But what if aggregate movements aren't in line with changes in any visible microdata? What happens, for example, when a report points to a particularly weak quarter for banking profits and ostensibly contradicts the headlines for record profits by the two biggest banks? What if the CPI indicates an unchanged price level and you hear daily reports for huge spikes in most supermarket prices? How do you judge cases when the macro appears hanging, and either not supported or contradicted by any available microdata?

You can rely on a couple of pointers here. First, we know that the movement of aggregates can be different than the movement of its parts. Some firms in an industry may do better than others. Some CPI prices may increase, some may decrease, and many will stay put from one period to the next. In such cases, the movement of aggregate statistics reflects both the *change* and the *relative influence* (weights) of their individual components. Even contradictory microdata information is not necessarily indicative of errors. Second, good data releases will endeavour to provide explanations. The increase in total industry revenue may be entirely due to a very good run by a few big firms, with many smaller ones suffering or even folding. The increase in food prices may be offset by a precipitous drop in gasoline prices due to OPEC (the organization of the petroleum exporting countries) infighting and oversupply. However, when no explanation is provided, when there's no micro evidence whatsoever to corroborate an alleged aggregate movement, or when any available microdata totally contradict the aggregate, it's fair to be suspicious. These may be cases of 'macro adjustments', giving aggregates a life of their own completely disconnected from the microdata. Particularly in aggregates with not enough history or from non-proven sources, it'd take blind trust to buy the concept, its magnitude, and now its movement without an inkling of evidence underneath. Such a leap of faith is not in the job description of a fact-checker. You need to call it out by raising questions. The next move will not be yours. A response may set the record straight, but silence too can be interpreted.

*Synthetic data:* The expanded interest in microdata for research has intensified thinking on approaches to protect privacy and confidentiality. Sometimes, anonymized public use microdata files stripped from identifiers are constructed. This is more applicable to households and individuals, since there's no good way for some businesses in some industries to pass incognito. Other times, fictitious data can be used by researchers for the early testing of regressions and other techniques. Modelled data, based on proxy relationships between variables, are also constructed when quality issues or lack of granularity are concerned. Synthetic data have appeared as an additional method to facilitate microdata research. Effectively, they're simulated data meant to stand-in for the real ones. The distinguishing feature behind this idea is to preserve the format and key relationships among the real data without disclosing them. For instance, the Bureau of the Census in the US has been working on such data for the American Community Survey (ACS), a popular source for studies.[13]

The synthetic data approach is subject to criticism, however. Which relationships to preserve among the many is not trivial. Even then, how to model the relationships in the real data and generate synthetic data is open to holes. Working with real data offers more possibilities to find better relationships in the data than those used for the production of synthetic files. A posting by David Swanson[14] uses what's known as Anscombe's quartet[15] (Table 8.3) to demonstrate how synthetic files can mislead. Each one of the four of example data, all with the same mean and standard deviation for X and Y, and all producing the same regression model ($y = 3 + .5x$, $R^2 = 0.67$), could have generated the same synthetic data set. However, set 1 shows a linear relationship, set 2 a non-linear one, set 3 contains an observation that could have been seen as an outlier in the real data, and set 4 indicates a non-existing relationship between X and Y (for which no model can construct a synthetic file).

Constructing synthetic files requires considerable effort. Unless they exist, creating such files may absorb more time than the actual research, and there's no guarantee that they'll perform as expected. Nevertheless, as we've always maintained, depending on the nature, the objectives, and the stage of maturity of research, there's a time and a place for different approaches, synthetic data included. Chances are that studies based on synthetic data will identify them as such and offer explanatory descriptions. Either way, keep an open eye.

*Machine learning data:* A related way to construct microdata is through automated machine algorithms that 'learn' as they're fed with real data. Such is the case, for example, of hundreds of millions of computer-generated building footprints by Microsoft's Bing Maps, initially in the US, followed

**TABLE 8.3**

Anscombe's quartet

| | Anscombe's Quartet | | | | | | | |
|---|---|---|---|---|---|---|---|---|
| | Data Set 1 | | Data Set 2 | | Data Set 3 | | Data Set 4 | |
| | X | Y | X | Y | X | Y | X | Y |
| | 10 | 8.04 | 10 | 9.14 | 10 | 7.46 | 8 | 6.58 |
| | 8 | 6.95 | 8 | 8.14 | 8 | 6.77 | 8 | 5.76 |
| | 13 | 7.58 | 13 | 8.74 | 13 | 12.74 | 8 | 7.71 |
| | 9 | 8.81 | 9 | 8.77 | 9 | 7.11 | 8 | 8.84 |
| | 11 | 8.33 | 11 | 9.26 | 11 | 7.81 | 8 | 8.47 |
| | 14 | 9.96 | 14 | 8.10 | 14 | 8.84 | 8 | 7.04 |
| | 6 | 7.24 | 6 | 6.13 | 6 | 6.08 | 8 | 5.25 |
| | 4 | 4.26 | 4 | 3.10 | 4 | 5.39 | 19 | 12.50 |
| | 12 | 10.84 | 12 | 9.13 | 12 | 8.15 | 8 | 5.56 |
| | 7 | 4.82 | 7 | 7.26 | 7 | 6.42 | 8 | 7.91 |
| | 5 | 5.68 | 5 | 4.74 | 5 | 5.73 | 8 | 6.89 |
| mean | 9.00 | 7.50 | 9.00 | 7.50 | 9.00 | 7.50 | 9.00 | 7.50 |
| std deviation | 3.32 | 2.03 | 3.32 | 2.03 | 3.32 | 2.03 | 3.32 | 2.03 |

by Canada, and expanded to Australia and countries in Africa and South America. Using satellite imagery, algorithms were employed to produce microdata through semantic segmentation (recognizing building pixels) and polygonization (converting pixels to polygons).[16] The microdata were released as open data. It was estimated that in 1,000 randomly sampled buildings from the entire output corpus the false positive ratio is less than 1%. However, best anyone can say, the quality of the data varies among places, urban and rural settings, mountains and plains. When the data are used in local applications, closer inspection would be needed.

Keep in mind that a lot of the microdata in such machine learning data sets are real. For example, in Canada building polygons were released by dozens of municipalities as open data. These data were compiled and used to train the machine learning algorithms. In case the quality is not deemed sufficient for some uses, remember not to throw out the baby with the bathwater.

## Notes

1 Mark Hoekstra, "The Effect of Attending the Flagship State University on Earnings: A Discontinuity-Based Approach," *The Review of Economics and Statistics*, MIT Press, 91, 4 (November 2019), pp. 717–724, https://citeseerx.ist.psu.edu/view doc/download?doi=10.1.1.297.2883&rep=rep1&type=pdf

2 Jonathan Smith, Joshua Goodman and Michael Hurwitz, "The Economic Impact of Access to Public Four-Year College," *Annenberg Brown University*, (EdWorkingPaper: No. 20-229), 2020. Retrieved from Annenberg Institute at Brown University: https://doi.org/10.26300/0hgf-4s95

3 Seth D. Zimmerman, "The Returns to College Admission for Academically Marginal Students," *Journal of Labor Economics* 32, 4 (October 2014), pp. 711–754. Retrieved from https://sites.google.com/site/sethdavidzimmerman/research

4 Ben Ost, Weixiang Pan and Douglas Webber, "The Returns to College Persistence for Marginal Students: Regression Discontinuity Evidence from University Dismissal Policies," No 9799, *IZA Discussion Papers, Institute of Labor Economics* (IZA), March 2016. https://docs.iza.org/dp9799.pdf

5 Seth Stephens-Davidowitz, *Everybody Lies: Big Data, New Data, and What the Internet Can Tell You About Who We Really Are*. New York: Harper Collins, 2017.

6 Atila Abdulkadiroglu, Joshua D. Angrist and Parag A. Pathak, "The Elite Illusion: Achievement Effects at Boston and New York Exam Schools," *Econometrica*, Econometric Society 82, 1 (January 2014), pp. 137–196.

7 In Canada, men born between 1963 and 1966 could be linked to a parent when they were between the ages of 16 and 19 years. Information on the children existed when they were between 15 and 31 years old. Thus, the oldest members of the cohort under study were 15 in 1978 and the youngest 31 in 1997. Miles Corak, Bjorn Gustafsson and Torun Osterberg, "Intergenerational Influences on the Receipt of Unemployment Insurance in Canada and Sweden," August 2000. Paper available at SSRN: http://dx.doi.org/10.2139/ssrn.250754

8 Daouda Sylla, "Enterprise Size Class Transitions in Canada," *Statistics Canada*, (Catalogue no. 18-001-X ISBN 978-0-660-25327-5), Release Date: March 15, 2018, www150.statcan.gc.ca/n1/pub/18-001-x/18-001-x2018001-eng.pdf

9 John P.A. Ioannidis, "Why Most Published Research Findings Are False," *PLOS Medicine* 2, 8 (August 2005), p. e124, doi:10.1371/journal.pmed.0020124

10 Erick H. Turner, Annette M. Matthews, Eftihia Linardatos, Robert A. Tell and Robert Rosenthal, "Selective Publication of Antidepressant Trials and Its Influence on Apparent Efficacy," *New England Journal of Medicine* 358 (2008) pp. 252–260, www.nejm.org/doi/pdf/10.1056/NEJMsa065779

11 Annelieke M. Roest, Peter de Jonge, Craig D. Williams, Ymkje Anna de Vries, Robert A. Schoevers and Erick H. Turner, "Reporting Bias in Clinical Trials Investigating the Efficacy of Second-Generation Antidepressants in the Treatment of Anxiety Disorders: A Report of 2 Meta-analyses," *JAMA Psychiatry* 72, 5 (2015), pp. 500–510, doi:10.1001/jamapsychiatry.2015.15, https://jamanetwork.com/journals/jamapsychiatry/fullarticle/2205839

12 Statistics Canada, "Canadian Statistics Advisory Council, 2021 Annual Report – Strengthening the foundation of our National Statistical System," Release Date: December 16, 2021, www.statcan.gc.ca/en/about/relevant/CSAC/report/annual2021

13 See United States Census Bureau, "What are Synthetic Data," Last Revised: November 18, 2021, www.census.gov/about/what/synthetic-data.html

14 David Swanson, "The Fundamental Flaw Underlying Synthetic Microdata: A Simple Example," *Population Association of America: Blog Viewer*, 2021, www.populationassociation.org/blogs/paa-web1/2021/06/07/the-fundamental-flaw-underlying-synthetic-microdat

15 F. J. Anscombe, "Graphs in Statistical Analysis," *The American Statistician* 27, 1 (February 1973), pp. 17–21, copyright © American Statistical Association, reprinted by permission of Taylor & Francis Ltd, www.tandfonline.com on behalf of American Statistical Association; www.sjsu.edu/faculty/gerstman/StatPrimer/anscombe1973.pdf

16 Microsoft Bing Maps, "Microsoft Building Footprints," www.microsoft.com/en-us/maps/building-footprints

# 9

## Data Analysis

Our inquisitive nature and our insatiable curiosity generate endless questions. From the existential *who are we* and *where are we going* to questions regarding the environmental, economic, and social arenas we have our minds constantly occupied and our hands full. Answers to early questions lead to more questions and new ones are added to the mix as everything evolves, practically guaranteeing that we'll never run out. We strive to understand *how* our world is but also *why* it is the way it is. Often times, particularly when it comes to socio-economic matters, we even have opinions on how things *should be* – and act accordingly. To explain our reality and answer our questions, we devise theories, test hypotheses, and are always on the lookout for patterns. From the Big Bang to the most mundane, we rely heavily on quantification. From the invention of number to our days, our quest for knowledge goes through numerical data.

The whole idea behind all the data we produce is to learn from them. Learning occurs through a set of activities that we bundle together under the umbrella term *data analysis*. Some activities involve simple processing of the data to extract relevant information by arriving at some totals, summary statistics, or more detailed frequency distributions. Other activities involve complex manipulations of the data through sophisticated econometric techniques typically aiming at finding causal relationships among variables of interest or quantifying the relative influence of multiple determinants that we suspect drive a certain phenomenon. Data analysis can have a cross-sectional focus, looking at the characteristics, status, or behaviour of different segments of a population at a given time. It can also be of a time-series nature interested in the direction, magnitude, and pace of change.

## A Marvellous Toolbox

The analysis of data has a variety of tools at its disposal to distil and communicate insights. Like the contents of an ordinary toolbox, many are well known and used routinely, taught to kids at early grades, and are a part of our daily vocabulary. Others are still intimidating to many and reserved for more specialized applications. They're all used to produce statistics that

DOI: 10.1201/9781003330806-9

crystallize the knowledge we extract from the data and help tremendously to spread it around. As a rule, the main function of these tools is to simplify the complexity hidden in the data.

Summary statistics for individual variables are commonly computed with measures of central tendency and dispersion. The mean, the median, and the mode convey succinctly useful information no matter how lengthy a data series may be. The arithmetic mean is the dominant notion of the *average* and its use is commonplace. Almost instinctively, we work out in our minds that the average of any two single-digit numbers adding up to ten is five (Table 9.1). The geometric mean is at times preferred in applications involving fluctuating or exponential growth or when we care about equal effects of changes across dimensions in composite measures, such that a 1% change in life expectancy has the same effect on the Human Development Index (HDI) as a 1% change in output. In daily life, though, eyebrows would be raised in finding out that the average of 9 and 1 is not 5 but 3 ($\sqrt{1 \times 9}$). Even more frowns would follow the 1.8 average computed by the harmonic mean in this case. Being the reciprocal of the arithmetic mean of the reciprocal values, $1/([1/9 + 1/1]/2)$, it's the lowest of the three. Still, it's better suited for some applications in finance, engineering, and elsewhere. For instance, travelling a distance of 200 km at 100 km/hr and coming back at 80 km/he the correct average speed is not 90 km/hr, as the arithmetic mean of the two speeds would indicate. It took 4 1/2 hrs for the 400 km travelled, two hours to go and two-and-a-half to return. Therefore, the average speed was 88.9 km/hr (400 km/4.5 hrs), as computed by the harmonic mean $1/([1/100 + 1/80]/2])$. When travelling at different speeds for the same amounts of time, the arithmetic mean would be the correct measure of the average speed.

Frequently, we also calculate *weighted averages* when some individual observations are deemed to matter more than others. This is the case, for instance, when we weigh changes in prices with the expenditure shares of goods and services in a basket, or changes in asset prices with their relative shares in a portfolio. In principle, averages should be presented together with

**TABLE 9.1**

Simple tools for data analysis

| | | Mean | | | | Standard |
|---|---|---|---|---|---|---|
| # 1 | # 2 | Arithmetic | Geometric | Harmonic | Range | Deviation |
| 1 | 9 | 5.0 | 3.0 | 1.8 | 8 | 4.0 |
| 2 | 8 | 5.0 | 4.0 | 3.2 | 6 | 3.0 |
| 3 | 7 | 5.0 | 4.6 | 4.2 | 4 | 2.0 |
| 4 | 6 | 5.0 | 4.9 | 4.8 | 2 | 1.0 |
| 5 | 5 | 5.0 | 5.0 | 5.0 | 0 | 0.0 |

measures of dispersion that convey additional information about the series. The pairs of numbers that all produce the same arithmetic mean in the simple example of Table 9.1 represent different patterns. These can be summarized by the *range*, which indicates the difference between the extreme values, and the *standard deviation* or *variance*, which measures the differences of all observations from the mean. In the 1 and 9 pair, one number is nine times larger than the other, the range is 8, and the standard deviation is 4 ($\sqrt{([(1-5)^2 + (9-5)^2]/2)}$). When the mean is 5, such value for a standard deviation is quite large, and we'd know that there are big differences in the numbers behind the mean even without knowing their values. Both the range and the standard deviation become smaller as the values of the pairs get closer, until they disappear when the numbers are identical (5 and 5).

In data analysis there's a crucial interplay between absolute and relative values. Both matter for a complete understanding of any issue at hand, and both are necessary for the communication of messages. Relative values are derived through unit-free measures, such as percentages, ratios, and rates of change. The use of percentages started centuries ago and is now ubiquitous. They conceal the complexity of large numbers and are quite useful in comparisons across scales with very different magnitudes. A statement that 80% of the fans in a hockey arena cheer for the home team comes across much smoother than saying 14,316 out of 17,895. However, the absolute value of the denominator is key to a fuller understanding and indispensable to subsequent analysis. For example, the percentage of individuals with a smartphone in Luxembourg may be higher than in India but their absolute number will be within the margin of error of the smartphone holders in India. If anything, because percentages are so convenient, sometimes we allow our momentum to carry us farther than warranted. When small numbers are involved, I much prefer saying that 3 in 4 shots hit the target or 7 out of 11 players in a soccer team are right-footed than saying 75% or 63.6%, respectively. The only real numbers that matter can be absorbed by all and communicated easily. In such cases, rather than an opportunity, percentages are rather awkward.

Rates of growth, or decline, are also extremely common and informative statistics. They capture both the direction and the pace of change over time, and can be applied to consecutive periods or periods far apart. Sometimes the transition of a variable from one period ($t_o$) to another ($t_n$) can be subject to turbulent fluctuations, with abrupt ups and downs. The annual rates of growth over the $t_o - t_n$ period will be very noisy. Then *compound growth rates*, an application of geometric means, provide a straight-line abstraction of such movement through equalized rates in each intervening period (like equal billing practices by utilities). For the most part, rates of change work quite well, summarizing complex movements whose description would otherwise be cumbersome and allow for comparisons across different variables. A glaring limitation, though, is that they don't work when negative values are concerned. For instance, growth rates for revenues and expenses are both useful and expected, but applying them to profits is not a productive idea – they will break down the first time a loss is involved. Profits and losses can be better expressed as a proportion of revenues.

Another inherent feature of growth rate calculations is that the same absolute change results in lower growth rates as the denominator becomes larger. Adding one more unit to one makes it grow by 100%, adding it to 100 represents a growth of 1%. Such relativities are behind the comment you may hear off-and-on that a year in the life of teenagers 'feels' much longer than in the life of seniors. This, of course, is a purely mathematical property but can at times turn controversial and lead to arguments, if not appropriately framed. Making too much out of exorbitant rates of growth in brand new phenomena or the penetration of new technological products can be misleading, particularly if not accompanied by the absolute numbers. The trajectory of new products typically follows an S-curve, exhibiting very rapid and accelerating growth early on (convex part), giving way to slower and decelerating growth after (concave part), before it reaches a plateau and flatlines near-saturation levels. The bottom line is that describing growth at low absolute levels, whether in a total population or subpopulations of interest, requires due diligence and judgement. At a minimum, it cries for additional data rather than growth rates alone, which may border the meaningless. Analogous is the case of runaway inflation, or hyperinflation. A change in the rate of inflation from 2% to 5% is consequential, and could lead to a change in the stance of monetary policy with all its implications. However, confronted with extraordinary measurements in the order of 7,000% and 13,000%, a meaningful interpretation is awfully hard to come by. Attempting to find meaning or implied actions in the 6,000% difference between the two is futile. Before anything else, such numbers really convey the message that the price system has effectively collapsed. There are no nominal prices that anyone should expect to see in the next visit to a grocery store.

Our toolbox also contains *index numbers*, which are additional unit-free and dimensionless statistics. Indices allow abstract comparisons of series with vastly different numerical values and enable comparisons across variables expressed in different units. The conversion of actual quantities, whatever they are, to *pure* numbers makes possible comparisons even among proverbial apples and oranges. The construction of indices is versatile. It requires the choice of a *base period*, which typically is assigned a value of 1 or 100 for easier calculation of growth. The base year can be the first in a series or any other. If the former, index numbers start together before they diverge; if the last year serves as the base, all series start from different points and end together; if a middle period is chosen as the base, the indices will cross at that point. For example, at first sight, the figures behind the curves in the left part of Figure 9.1 are not terribly helpful to discriminate between the movement of a select small group (A) and that of the population at large (D). It's also difficult to compare the movements of

**FIGURE 9.1**

What are index numbers good for

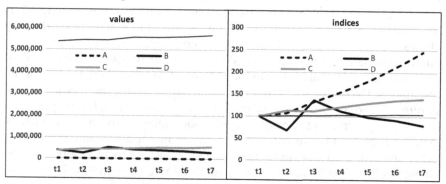

other groups (B and C), both between themselves and against the others. Converting them to indices and plotting them with a common start offers a starker view of their relative evolutions.

Such methods mostly serve single variables. Cross-tabs are another technique that makes it possible to look at data across multiple dimensions. They reveal relationships within a data set that aren't apparent from any one dimension alone, and exemplify granularity.

Data analysis can be fun and is sometimes perceived as a reward for the efforts to generate the data. However, while familiarity with the production of data is definitely an asset, analysis is fundamentally a different activity with its own set of skills. How well can we use the data to distil messages is up to us. The same data can be used or abused. Whether data are in short supply or bountiful, deriving analytical insights is higher in the pecking order of knowledge generation. Learning from analysis is incremental and leads to the formulation of better and higher-order questions. Some of the answers we seek through analysis are actually different configurations of the data themselves, others need to be inferred indirectly. A broad distinction is between *descriptive* and *inferential* analysis, for which an entire apparatus of tools and a great number of techniques have been developed.

## Descriptive and Inferential Analysis

Descriptive analysis can paint an accurate picture of anything we want to know. Using data to illuminate the magnitude and the make-up of populations of interest represents a powerful and indispensable first step before any other investigative activity. The more granular the data, the more detailed the picture will be. Data would be worth a lot even if that's all they were good for. Descriptive analysis reveals the 'lay of the land'.

In fact, a good analogy for what descriptive analysis can accomplish is the creation of a map. Starting with oceans and land masses, a world map allows us to recognize continents and their shapes, measure their extent, and their distance from each other. With more data we can add rivers, lakes, and mountains. We can measure the lengths of rivers and the heights of the mountains. The discovery of America, polar explorations, and missions to the rainforests and jungles can be seen as data expeditions too, for the purposes of mapping. Increasingly, maps become more detailed and refined, particularly with better instrumentation. We can map countries and their cities, every street in a town, every house on a street, every room in a house. The OpenStreetMap[1] and other initiatives are still going strong today. The level of detail has really no end. We can map every tree in the forest or every leaf on a tree, if we wanted. But at some point, for most of our purposes, we feel sufficiently equipped to navigate with confidence – with the added proviso that we pinpoint our own coordinates on the map. The map created by our data becomes then quite prescriptive. Alternative paths for moving from A to B can be plotted and assessed. River crossings are visible, unfriendly terrain is red-flagged, and we can choose a scenic route rather than a highway. Similar processes transpire with statistical data used in descriptive analysis.

Answers to our early questions provide short-lived satisfaction before they fuel our next line of relentless questioning. Knowing how many continents and mountains there are becomes the springboard for more ambitious queries. We're itching to know if the continents move, how were the mountains formed and why some are taller, how are water levels affected by the phases of the moon, and a myriad such things. Repeated measurements and descriptive analyses over time can answer more questions but also trigger new ones. Accuracy becomes an issue, particularly for small differences. We need to have confidence that Africa edges towards Europe by 2 cm a year before attempting any grandiose explanation or, worse, start preparations for impact. We'll always have in the back of our minds the possibility of measurement errors as the potential of descriptive analysis is exhausted and inferential methods take over. Clearly, the more solid descriptive analysis is, the better equipped we'll be for inferential analysis.

Descriptive statistical analysis is deployed throughout our socio-economic inquiries. Pegging the population of a country at 50 million in 2020 leads to a series of questions. Who are these people, where do they live, what do they do? Knowing their age distribution alone is very useful to governments and businesses. Even more so is knowing how it's changing. In the example of a plausible society in Table 9.2, the population grew at a healthy rate since 2000. It grew faster for people 50 years and over and, by consequence, the relative shares of those age groups in the total population also grew (shaded cells). The absolute number of people in each age group between 60 and 89 more

**TABLE 9.2**

Detailed data can be quite revealing

| Age | Population | | Distribution (%) | | Growth (%) | Difference | % |
|---|---|---|---|---|---|---|---|
| | 2000 | 2020 | 2000 | 2020 | | | |
| 0–9 | 3,827,648 | 5,218,422 | 12.5 | 10.4 | 36.3 | 1,390,774 | 7.2 |
| 10–19 | 4,151,752 | 5,489,987 | 13.5 | 11.0 | 32.2 | 1,338,235 | 6.9 |
| 20–29 | 4,142,589 | 6,746,494 | 13.5 | 13.5 | 62.9 | 2,603,905 | 13.5 |
| 30–39 | 4,942,224 | 6,962,728 | 16.1 | 13.9 | 40.9 | 2,020,504 | 10.5 |
| 40–49 | 4,916,363 | 6,386,439 | 16.0 | 12.8 | 29.9 | 1,470,076 | 7.6 |
| 50–59 | 3,600,687 | 6,834,336 | 11.7 | 13.7 | 89.8 | 3,233,649 | 16.7 |
| 60–69 | 2,395,189 | 6,219,558 | 7.8 | 12.4 | 159.7 | 3,824,369 | 19.8 |
| 70–79 | 1,809,938 | 3,953,304 | 5.9 | 7.9 | 118.4 | 2,143,366 | 11.1 |
| 80–89 | 770,795 | 1,748,548 | 2.5 | 3.5 | 126.8 | 977,753 | 5.1 |
| 90–99 | 125,302 | 425,032 | 0.4 | 0.9 | 239.2 | 299,730 | 1.6 |
| >100 | 3,243 | 15,152 | 0.0 | 0.0 | 367.2 | 11,909 | 0.1 |
| Total | 30,685,730 | 50,000,000 | 100.0* | 100.0 | 62.9 | 19,314,270 | 100.0* |
| Median | 37 | 41 | | | | | |

Note: *Some totals may not add up due to rounding.

than doubled, in the 90–99 age group more than tripled, while the number of centenarians more than quadrupled to a historic high exceeding 15,000. If we wanted, we could distribute the increase in the population between 2000 and 2020 by age group too, as in the last two columns. The age groups of people 50 and over accounted for more than half the total increase in the population, with almost 20% coming from the 60–69 age group alone. The population under the age of 50 still accounts for more than 60% of the total but that dropped from over 70% in 2000. Overall, the country's population grew older with a median age of 41, up from 37 twenty years ago.

Even from such simple descriptive analysis we learn a lot about ourselves. Still, many questions remain unanswered. How many births and deaths occurred during that period? What was the net migration? Was the transition smooth over the years or did a few years account for most of the change? Surely, additional data and more detailed descriptive analysis can answer all such questions too. We'd learn even more if we looked at a much bigger and detailed spreadsheet with annual data, for every age, across more dimensions. But we'll certainly need all the microdata available for the next order of questions involving inferential analysis. What are really the forces behind the population movements shown by the data? What affects fertility, mortality, and migration? What are the implications of a higher median age? Does diet or hygiene contribute more to longevity? Results from such analysis will be explanatory in nature, often interesting and, at times, tentative and controversial. They will certainly be less factual. Also, imagine the questions if we were in the year 2000 and the 2020 data were projections to help planning. This too is in the realm of inference.

**TABLE 9.3**

Distribution of households by type from 1960 to 2020

| Households | Distribution of Households (%) | | | |
|---|---|---|---|---|
| | 1960 | 1980 | 2000 | 2020 |
| Married couples with children | 45 | 35 | 28 | 20 |
| Married couples without children | 30 | 30 | 30 | 30 |
| Single parents with children | 4 | 6 | 8 | 10 |
| Other family households | 5 | 5 | 6 | 5 |
| One-person households | 10 | 18 | 25 | 32 |
| Other non-family households | 6 | 6 | 3 | 3 |
| Total | 100 | 100 | 100 | 100 |

Descriptive and inferential analysis feed on each other continuously as we keep asking questions and learning more about ourselves. When politicians battle it out for our vote, interest in families is all the rage. An observer would get the distinct impression that heterosexual married couples with kids are the cornerstone of our societies. Looking at statistics, though, we see that this stereotype is heavily influenced by times gone by. Descriptive analysis based on our constructed notions of *household* and *family* shows how family pronouncements really stack up. While the exact situation differs among countries, the stylized example in Table 9.3 speaks well to the evolution of the population structure in modern western economies. The coveted traditional family, which accounted for 45% of households back in 1960, represents only one-fifth of households in 2020. In Canada, in the early part of the 2000s, it fell for the first time below the proportion of married couples with no kids, which has been remarkably stable over the decades. Moreover, the ongoing increase in one-person households put them ahead of both by 2020. This has been the case in most western countries, where one-person households moved from the single digits half a century ago to the largest category today. In the US, the UK, France, Sweden, and Japan their proportion hovers around 30% or is higher. In Norway and Germany it even exceeds 40%. Looked at another way, in 2000, in Canada this group represented about 15% of the population over the age of 15, compared with barely 2% half a century ago. Single-parent families also have grown significantly and attract policy attention. Moreover, about eight in ten such families are headed by women.

Figure 9.2 shows graphically the evolution of some key groups. Their ending points in 2020 are quite different from their starting points in 1960. The insert displays the trajectories of families versus other households which, in addition to single individuals, include living arrangements with roommates.

There are a lot more descriptive statistics (not shown) that embellish the same story. Over the years, and as populations grew, the once large average household size in the western world has plummeted. In many countries it's now between 2 and 2.5 members. The marriage rate has fallen, and marriages

**FIGURE 9.2**

The evolution of household types between 1960 and 2020

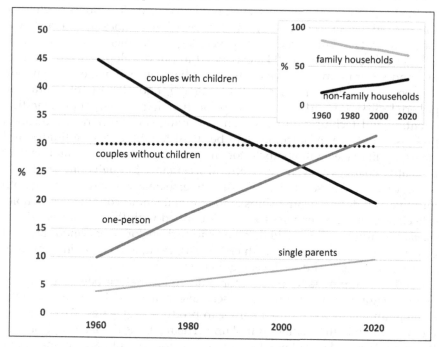

take place at higher ages. Childbearing has also been pushed for later, and birth rates have fallen below the replacement rate of 2.1 – in many countries, way below. Still, couples (with kids or without) account for more than half the households and, in some countries, two-thirds. However, not all couples are married today. Common law has increased significantly accounting for approximately 20% of all couples, 3–4 times more than half a century ago. Same-sex couples are also counted now and are growing. Many complex households have been formed, comprising blended families with step-parents and siblings.

Through a series of snapshots, descriptive statistics can reveal even more details with the use of cross-tabs. In single individuals, the relative shares of the unattached young will be contrasted with widows. Couples without kids will be split into childless couples and empty nesters. We'll even find out that in South Korea the proportion of married couples where the female is older than the male has increased from 11% to 16% in the last two decades. At the same time, we're moving beyond a binary gender choice.

Even before all the details are mapped, nagging questions will preoccupy our minds. The baton will be passed to inferential analysis. Why's all this happening? What forces are behind these changes? Which of them exerts the most influence? Will these trends accelerate, stabilize, or reverse? What is their deeper meaning, and what are we supposed to do? Now is the time for the big theories, the models, even our conjectures. The doors open wide and the red carpet is rolled out for all sorts of disciplines to take a shot at the answers. Economics, sociology, psychology, and other behavioural fields will get to work. Efforts will be expended to put developments in perspective and separate main movements from aberrant or transitory behaviour. Demographic shifts, family structures, ageing, and all sorts of interaction patterns will be examined in the context of changing economic opportunities, scientific and technical advances, and evolving societal norms. Education processes and health outcomes will be dissected. Studies of attitudes and emerging lifestyles will compete for the winning rationale for and against legislative changes or even vie for ethical acceptance through numbers. Every aspect of life will be scrutinized, mixing diverse data sets and exploring both intuitive and daring relationships among them. Under conditions of wholesale change, high-order forces will be identified with the use of *other things equal* assumptions. These will then become anchors for an enormous amount of very specialized inquiries with claims left, right, and centre in a multidimensional context. And we'll keep going.

Detailed mapping takes place before inferential analysis begins. Ironically, deeper digging under the guise of descriptive analysis makes more use of the same granular microdata used earlier to produce aggregates. We can go all the way down to the last individual unit. The relationship of an analyst with the microdata is like the knowledge a good leader has of his team members. It goes past their professional qualifications, strengths, and weaknesses and extends to family matters and intimate personal traits. Analytical work also exploits a critical link, the feedback mechanism from inferential to descriptive analysis. Empirical uses of the data represent the best hope not only to catch any data errors but also to identify data gaps. In its broad sense, inferential analysis extends well beyond its original application of inferring truths about a population based on a sample and encapsulates all the heavy lifting in research. The attempts to explain why things are the way they are and why they're changing, to quantify the relative magnitude of different forces, to identify cause and effect, or to project future paths are all under the purview of the notion of inference. In its cause, many methodological approaches are employed, including regression analysis and hypothesis testing.

## Regressions and Correlations

Elections are decided by individual votes. The decision-making behind each vote can be quite different. Some votes are cast based on ideology and allegiance to a party over many decades, regardless of 'details'. Others may be based on detailed comparisons of party platforms across multiple issues. Some individuals may be one-issue voters. At times, local influences may be the determining factors. Simply liking or disliking the face of a candidate is not out of the question. The end vote somehow encapsulates all the above and more in a complex weighting scheme. Our data points are also the products of multiple forces at work, frequently colliding and bumping against each other, tying into knots the relationships among variables. In the real world, simply put, too much is happening simultaneously. This in no way discourages us from trying to untangle the underlying complexity, frequently one string at a time. The *ceteris paribus* assumptions are only the beginning of trusting our 'rational' reasoning. To reduce poverty, eliminate the budget deficit, or achieve some other societal outcome we need to know what actions to take, with what intensity, and when to stop. Even without overly ambitious aspirations for fine-tuning through minute calibration of dosages, enough knowledge of how therapies work is needed. For that we devise theories to guide us, postulate relationships among variables, and assess how well they work.

Regression analysis is used to estimate the relationship between a dependent variable (Y) and one or more independent variables (Xs), also known as *explanatory variables* or *covariates*. Typically, they're represented by an equation, including an intercept ($\alpha$) and a residual or error term ($\varepsilon$).

$$Y = \alpha + \beta_1 X_1 + \beta_2 X_2 + \ldots \beta_n X_n + \varepsilon$$

How the specific equation to be tested is arrived at can be a long story – or a very short one. Sometimes it's the *reduced form* of a whole *structural system of equations* with a lot of theory behind it. Other times it's just the best specification of determinants a researcher could think of. In univariate or multivariate regressions, the coefficients ($\beta$s) purport to reveal the relative influences of the independent variables on the variable of interest. The unexplained part is absorbed by the error term, which may also account for missing variables. When theory is involved, like in *growth accounting*, the output growth is decomposed to the contributions of the factors of production, capital, and labour. The remainder is captured in the *residual* and is interpreted as productivity gains from all factors or the effects of technological change. When data substitute for theory, analysis of the error term will look at autocorrelation to ascertain if there's a confounding factor. If only *white noise* is found there, the researcher may feel that the model performs decently.

Regressions involve a fair amount of data manipulations. The variables can be expressed in a variety of ways, such as first differences, rates of growth, or logarithmic transformations. Many variants of regressions can also be set up depending on the type of data used, the exact technique, and the objective of the exercise. In their beginning, they were only linear. Particularly when used in projections, they were subject to all the limitations that interpolation and extrapolation of straight lines have over long periods. The former predicts within the data set, the latter extends past it – if pushed far enough, based on the broken world records in the 100-m dash, it'll show that at some point some human will run the distance in negative time. Today, there are many techniques for non-linear regressions too. Also, when easier computation methods replaced manual estimation, non-parametric techniques for data analysis were developed, such as the *bootstrap* that draws model-free inferences through continuous resampling. Nowadays, a long menu of options exists covering cross-sectional and time-series regression models, continuous or discrete data sets, and series that exhibit various specific properties. Even *dummy variables* are used to separate the data by some categorical characteristic, like male/female or yes/no, affecting the intercept and incorporating qualitative elements in our quantities.

A mechanical function of regressions is that whatever two series are run against each other, the procedure will spit out a result every time. When input data are entered, the machine is agnostic to the meaning behind the numbers. It won't feel any 'quality', and couldn't care less if the numbers are meant to represent the GDP, the number of trays in a cafeteria, or anything at all. We won't hear back that the weekend attendance of movie theatres over the last two months has absolutely no relation to the number of llamas grazing around the Andes. In case you wonder, there's no need for anyone to run any regression simply to examine possible relationships between variables. Other inferential analyses or simple cross-tabs can easily reveal correlations among data series. Some will be stronger than others, whether the series move in the same or opposite direction. Moreover, some correlations will conform to our expectations based on theories, intuition, or conjecture. Others will jump randomly out of the data, puzzle us, or make us giggle. The example of the matching movements between the number of storks and the delivery of babies in some locations stick to mind. Or it could be the number of sociology doctorates awarded in the US and the number of deaths caused by anticoagulants, between 1999 and 2009. Given the volumes of data, there will be endless non-sensical correlations, even eerie or spooky at times. There are websites dedicated to them.[2] For research purposes, such correlations are worth a dime a dozen. On the flip side, you can easily see the makings of conspiracy theories by those prone, especially with a couple of somewhat plausible examples. One way to debunk faulty correlations is to extend the time periods involved, as most are carefully selected for effect. Some periods are truncated, conveniently or accidentally. This, in fact, can be a general issue in research design with no predefined ending. The more the data are extended before or after the stated periods, the higher the chances that any apparent correlation will break down. By contrast, true correlations will become stronger over time, as the irregular bumps that keep the variables temporarily apart will be smoothed out.

Except for the occasional utter nonsense, and beyond fake correlations, there are also spurious correlations. These occur when two highly correlated variables are, unknowingly, both caused by a third one. For instance, consumption of ice cream and taking cold showers may correlate well because they're both caused by unusually hot days. It'd be bizarre to think that one causes the other. But the presence of third factors doesn't necessarily imply spuriousness. There would be little questioning of the relationship between the increased use of air conditioning and electricity consumption, both of which are also caused by the same hot days. In any event, with correlation as an intermediate step, we get closer to our ultimate research interest, which is causation. For a long time now, much has been written on this topic and many warnings have been issued not to confuse correlation with causation. Our overall numeracy has benefitted from such exposure, and there's no need for redundant repetition here. In the final analysis, the issue really is how to establish the validity behind supposed relationships, something that must pass through hypothesis testing.

## Hypothesis Testing

As they go about their business, researchers frequently use real-world observations to build hypotheses. Then, they resort to further data gathering to see how such hypotheses fare. They hang on to every bit of corroborating evidence that adds credence to their hypothesis. It feels good to have the facts on your side. Trying to verify a theory with data, though, is not exactly how research works. In the philosophy of science, such an inductive approach to reasoning has taken a beating and a material nuance has been introduced. Theory can never be proven, goes the argument, it can only be falsified – by the first piece of evidence against it. Conceivably, a researcher may shrug off or explain away an isolated piece of contrarian evidence. It could perhaps be an error or a weird anomaly. Occasionally, even 'the exception that justifies the rule' card can be invoked in good humour. Regardless, a well-known argument is that relying on a pattern exhibited by the data is not sufficient to prove a theory, never mind how systematic and extensive the corroborating evidence may be. Best we can do is to treat it as correct, if we so choose, and keep testing it until conclusively disproven by new data! All the daily observations that the sun rises in the East and sets in the West every day for millennia cannot be trusted as sufficient 'proof' that it will do so tomorrow again. Under this logic, sunrises remain a useful theory – until the day the sun doesn't rise and it's disproven. Some may take issue with this particular example presented by Popper, myself included.* But, then, no one would seriously argue that a handful of corroborating data points constitute an absolute proof for a hypothesis. The breakdown of many correlations we discussed would lead to unbearable disappointment and loss of credibility sooner than later. How many data points are enough? Is guessing the drawing of an ace from a deck of cards enough to call someone a seer? How about two or three consecutive guesses? The

---

* Yet scientists point out that our solar system is unstable. At some point, in the very distant future, the sun will not rise.

influence of the thinking that the up-to-a-point absence of evidence against something should not be interpreted as proof for the presence of evidence for its opposite has been profound in statistical inference. It opened the door for probabilistic thinking to enter the scene.

The acceptance that we can never truly prove a hypothesis, only reject one in a probabilistic sense, has led us down the path of a double negative. We treat a relationship that we suspect to exist as the *alternative hypothesis* $(H_1)$, X causes Y, and formulate a *null hypothesis* $(H_o)$ that states in the negative that X doesn't cause Y. In a sense, the null hypothesis is a technical necessity and implies that X has no more explanatory power on the movement of Y than chance alone. The null hypothesis effectively plays the role of the devil's advocate, the nay-sayer. To anything we think may be happening, it takes a contrarian position. Suspecting that X causes Y, the null will state that X doesn't cause Y. Hoping that a new drug will help, it'll state that the drug isn't better than a placebo. Then, the whole idea is to use our data and determine the probability with which we can reject the null hypothesis, if not actually true, opening the door for the alternative to be valid. In that sense, while we can never prove that X causes Y, we can fail to prove that there's no chance that X might cause Y.

Figure 9.3 lays out the possible outcomes of this thinking. If the null hypothesis is true, we obviously make the correct decision when we don't reject it (sometimes referred to as we 'accept' it). But if we reject it, we commit type I error. This false positive result means that we give credence to the alternative, X causes Y, when none is deserved. Then, we'll continue to look for answers where they don't exist, similar to looking for our keys under the lamp post – any keys found would be the wrong set as ours were lost elsewhere. Conversely, if the null hypothesis is false but we don't reject it, we commit type II error (false negative). Effectively, we screen the true answer out prematurely and draw the wrong conclusion, giving up ever finding that X really causes Y. But if we reject the null hypothesis when it's false, we make the right decision again. The closest we can possibly come to 'proving' that X causes Y is to prove that we cannot prove that X does not cause Y!

Still, none of that is done with absolute certainty. Even when our data results show enough support for rejecting a false null hypothesis and start believing the alternative, how much is enough? That's when a probability threshold is established, under which what's observed by the data could happen by

FIGURE 9.3

Decisions around the null hypothesis in hypothesis testing

| Null hypothesis | don't reject | reject |
|---|---|---|
| true | correct decision<br>X doesn't cause Y | false positive<br>type I error |
| false | false negative<br>type II error | correct decision<br>X could cause Y |

chance alone. Enter the *p*-value – that we encountered in Chapter 8. Typically, a choice of 1%, 5%, or 10% is made as the *level of confidence* and every time the test returns *p*-values lower than those, the results are interpreted as indicative of *statistical significance*. The null hypothesis can be rejected within that level of confidence – what's contemplated is unlikely to happen simply by chance. This is no proof that the alternative hypothesis is correct, and the alternative can be whatever, really! If it were a *p*-value, the probability that someone successfully guesses the drawing of four consecutive aces from a deck of cards would easily produce 'statistically significant' results. Whether he'll be called a seer will always require a judgement call.

The reasoning behind *p*-testing has filtered through many practices in the real world. In the court system, there's no such thing as proving innocence directly. The onus is on proving guilt. Not proving it, with criteria such as the preponderance of evidence or beyond reasonable doubt (equivalent to different p-values), results in a *not guilty* verdict. This is as close to exoneration from a false accusation as an innocent defendant can get. Similar happenings occur in other situations, from health to education to the labour market or even mating. There's always uncertainty guiding our decisions, as if we test hypotheses. In employee hiring, for instance, a false positive will lead to the hiring of a bad candidate (type I error) and a false negative (type II error) to missing out on a good candidate. In mating, a false positive may lead to divorce and a false negative to missing the chance for the real soulmate. Errors are inevitable but systems will be successful or not based on their ability to correct them.

The excessive preoccupation with *p*-values has come under attack and has attracted many critics. Alternatives are tried, including some based on *conditional probabilities*. For example, the probability that someone who tested positive for COVID-19 is actually infected depends on the accuracy of the test but also on the prevalence of the virus in an area.[†]

The probability of testing positive depends both on false positives and false negatives. With prevalence at 10%, test accuracy at 95%, and interpreting | as *given that someone is* (or *if*), the probability of an actual infection following a positive test is mathematically expressed as:

$P(A|B) = P(B|A) \times P(A)/P(B)$ = accuracy × prevalence/P(positive) = (95% × 10%)/(14%) = 67.9% with,

- $P(B|A)$ probability of testing positive if infected = 95%
- $P(A)$ probability of being infected = 10%
- $P(B)$ probability of testing positive = P(true positive) × P(infected) + P(false negative) × P(healthy) = (95% × 10%) + (5% × 90%) = 14%

---

[†] Generally, test accuracy comprises *sensitivity*, the probability of testing positive when actually sick, and *specificity*, the probability of testing negative when healthy. In other applications they're also known as *power* and *reliability* and, typically, the two values are not the same.

**FIGURE 9.4**

Probabilities of being infected following testing

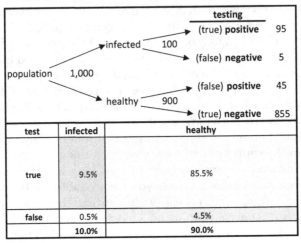

| test | infected | healthy |
|------|----------|---------|
| true | 9.5% | 85.5% |
| false | 0.5% | 4.5% |
| | 10.0% | 90.0% |

A more intuitive way to look at this is shown in Figure 9.4. The 10% *prior prob- ability* of being infected (prevalence rate) splits a population of 1,000 in 100 infected and 900 healthy individuals. If they're all tested, the 95% accuracy of the test further splits each of these groups. Taking into account both false positives and false negatives, the number of individuals testing positive is 140 (95 + 45). Therefore, the probability of someone with a positive test being actually infected with COVID-19 becomes 67.9% (95/[95 + 45]). The probability of being healthy despite a negative test is 99.4% (885/[885 + 5]). Such *posterior* probabilities will keep changing as more data accumulate. For instance, changing prevalence to 5% would change the posterior probability of a true COVID-19 infection after a positive test to 50%, and of being healthy after a negative test to 99.7%. Again, however, there can never be 100% cer- tainty as to whether someone is infected or not under any test result. Similar cases are encountered in doping and other applications. There will always be some lingering doubt.

## The Truth, the Fractional Truth, and Nothing But Enough Truth

Living in a world filled with uncertainty, it feels comforting to grasp onto anything stable. We place a premium on our ability to separate reality from fiction and tell truth from falsehood. At least since the days of the ancient Greek philosophers, we have formalized in logic that something can be either true or false. It can't be both simultaneously. What are we then supposed to make of our conclusion that a hypothesis is true with a 90% probability? Considering the probabilistic determination of inferential findings, we'd be remiss not to devote a few paragraphs on the notion of probability and its relationship to logic, truth, and even reality. Branches of philosophy continue to grapple with such questions.

For the most part, we can separate true from false just fine. But at times we slip, either because our own senses let us down, we allow our minds to play games on us, or fall prey to gaslighting. Realizing that we may be imperfect witnesses of what we saw and heard, we rely on video and DNA forensics to avoid or correct mishaps. We're masters of abstraction, yet we manage to trip on our own creations. We invented numbers and use them as objective markers to understand our world through quantification. Before long, though, we questioned whether we can really move from 1 to 2. An extra 9 can always be added to 1.999, forever approaching but never quite reaching 2. The same happens between 3.4 and 3.5. Infinities within endless infinities somehow make us fidgety. We stumble on whether 0.33333 is really $1/3$.[3] We've even managed to sow doubt on the existence of motion. To go from A to B, first we need to walk half the distance, then half of the remaining half ($1/4$), and so on. There will always be something left separating us from the destination. Or, to walk the first half of the distance we must first walk the first half of that ($1/4$) and, even before, the first half of that half ($1/8$), and so on. Again, there will always be a finite distance preventing us from starting to walk. So, either we can't finish or we can't start. This was just one of Zeno's paradoxes when he was playing devil's advocate. But he had more.

At an instant in time, a flying arrow occupies a space in the air equal to its length. During that instant, it can't possibly come from somewhere or go anywhere. Since time is made up by such instances, the arrow really stands still and motion is an illusion. Instances with no duration, points without length, and the like can trick our minds as we try to gain a deeper understanding of reality through 'proofs'. This can be healthy up to a point, particularly after working hours, lest our brains atrophy. Thankfully, experiential data eventually dominate and we snap out of that trench. We know well that water will put out a fire every time. We'll agonize later how this is possible since both hydrogen and oxygen make fire worse. Truth is we can easily move from 2 to 3 and effortlessly walk from A to B, as Diogenes the Cynic showed by just getting up and doing it. We can use modern differential calculus to explain motion, if we prefer. We can also use *reductio ad absurdum*, refuting a proposition by showing that its acceptance would lead to logical absurdities.

Analogous mental tripping happens in inferential analysis. There's no argument that the bar must be set high when we try to prove a hypothesis. We can't just allow a bunch of favourable data points to impersonate the truth, only to have it turned upside down with a few extra data points later. But, unlike the physical world, the 'truths' in the socio-economic domain are qualitatively different. Our underlying factual evidence, our data, capture the confluence of an enormous number of variables none of which can be expected to remain invariant. This will inevitably produce different outcomes than universal and irrefutable truths. For any hypothesis, our data will be commonly distributed in a way that contain evidence for and against. It takes sifting through all the evidence to assess what's the rule and what's the exception. Say, for example, that we wish to check the veracity of a company's claim for same-day delivery. We gather data and look at the frequency with which this particular promise holds true. But, even if the thousands upon thousands of data points that we may have all confirm same-day delivery, what kind of 'truth' is this? Inclement weather may get on the way, as can labour strikes, volcano eruptions or other natural disasters that delay flights and disrupt supply chains, power outages that wreak havoc in transportation, people who suddenly get sick, machines acting up, a flat tire, or anything else. There's no end to what can go wrong at some point for same-day delivery to fail. We don't know when and why, but it will. Best we can do is to say that it holds true, up to now.

This is precisely the point of the presence of randomness in life, and the unpredictability that lies behind the *black swan* thinking.[4] When all our existing data only show white swans, it can become our truth – until it's not because a black swan is spotted. It may be comforting to believe that things will continue to unfold as known in the past – until they don't. Here's why we rely on statistics and probabilities, based on frequencies of actual happenings rather than being fixed. Probabilities for car insurance come from accident statistics by age, for heart surgeries from the accumulated experience of cases over the years, the odds for betting in games of sport from past performance and present information about the teams. All these will be changing with more data.

To be fair, probability is not exactly an intuitive concept. We create a new imaginary world. While it can be useful, it shouldn't be confused with *fractional truth*. Such an intermediary level between truth and falsehood doesn't exist. The expected value of a die roll, a probability, is 3.5. It will never happen. It's 'true' within its imaginary, made-up world. Not in ours. Like a dream within a dream. Waking up from the second dream, we're caught inside the first, not the real world. When we have faith that the efficacy of a COVID-19 vaccine is 90%, the statement "*All those vaccinated against COVID-19 will not be infected*" is not true. Neither is the statement 90%, or fractionally, true. What is true is that "*90% of all those vaccinated against COVID-19 will not be infected*".

While past data are not enough to guarantee truth and predict the future, the socio-economic realm is distinct from the physical. Can we really extend the imaginary world of probabilistic inferences to sunrises, by choosing to see them as a testable hypothesis? We can't stop someone from that kind of thinking, but practically it matters little. Our data will keep rejecting the null hypothesis every day, without ever committing to absolute certainty for tomorrow. Now, not only this sounds like a bad game but this unfortunate example of the sunrise is too defensive a posturing and a step too far. We don't 'only' have to rely on millennia of daily data, we know so much more.

We've mapped the workings of the solar system and the movement of celestial objects. Isn't it best to see sunrises as reliable data points and combine them with many more and prove something bigger, such as the Bing Bang theory? Either way, it's not critical, the data about our sunrises won't change. All that matters is to resist data contamination from gaslighting, a prank that would point to midnight when the sun is noon-high. The sunrise example is a reductio ad absurdum in and of itself but, if more is needed, its 'logic' leads straight to another reductio ad absurdum – it's reminiscent of that we may not have enough data that death is inevitable. In any event, Godel's *incompleteness theorem* proved that even mathematics can contain 'truths' that can't be proven using only the axioms of a system. (Think of *proving* that something cannot be *proven*!) What does this say for Euclides who, millennia ago, postulated axioms to prove theorems? It's not my personal opinion that his geometry has served us exceptionally well regardless of whether its theorems were actually proven or thought to have been proven, and despite the non-existent abstract points that define straight and parallel lines or the more recent allure of Riemannian or elliptical geometry.

Outside of our mental games, other than wasting time and energy, such mental acrobatics frequently blur the real target. Our lived experiences inspire enough confidence in judgement. Throwing a stone and breaking a glass window every time shouldn't be that confusing in practice. Changing income tax rates will immediately move individual disposable incomes in the opposite direction, when disposable incomes are defined as after-tax incomes and deductions are applied at the source. This fact is distinct from the effect of disposable incomes on consumption later, a relationship that may hold at the aggregate level only. In a nutshell, agony to prove theoretical truths shouldn't be confused by the practicality of establishing facts, as our measurements do.

A corollary of all the above is that our descriptive data are harder, more conducive to defining reality than data from inferential analysis. The numbers that underpin descriptive analyses are definitive and authoritative. Assuming they're error-free, the numbers are what they are. There can't be any dispute or argument. Those that come out of inferential analysis, in the form of results, are not so. They're more conjectural. They're prone to change with the slightest variation in the specification of a regression or other technique that produced them. They're a different genre of data, situated in that intermediate world between reality and fiction that we've created and which exists only in our minds. This also explains the emphasis on the reproducibility of results. Only then, credibility is achieved. Inferential disagreements on the exact influence exerted by one force or another, the ceteris paribus assumptions, or something else are all fine – too many things are happening at the same time. Reasonable people can disagree but *"you're entitled to your own opinion, not your own facts"* applies. Descriptive statistics don't go the full distance and don't satisfy all our curiosities. But they go far enough for most practical purposes. They establish the foundations of our evidentiary reality quite well and do so much better than anything else we have. They don't contain all the truth but they contain enough truth.

## Probability and Human Nature

In principle, probabilities facilitate decision-making as they fit hand-in-glove with our attempts to peel off layers of uncertainty. However, the dry and precise answers that result from the mathematics of probability don't always register as neutral in the wiring of our brains. Behavioural research has shown amply that the implied 'logic' of probability is not the dominant factor either in understanding choices facing us or in subsequent decision-making. This is particularly so when we move away from games of chance with fixed probabilities to real-life situations with more fluid Bayesian *degrees of belief*, which will be modified with new data. The notion of probability has generated much wonderment and different interpretations among economists, statisticians, philosophers, and other thinkers over the decades.

We understand that under controlled circumstances a rare event, say, one in almost 6.5 million, is bound to happen. Such is the probability of drawing any hand of four cards from a deck of 52, whether the result is four consecutive aces or a hand containing a 2 and a 9 of spades, a king of hearts, and a 10 of diamonds. If we had to guess a given hand beforehand, the probability would be $(1/52) \times (1/51) \times (1/50) \times (1/49) = 1/6{,}497{,}400$ – without replacement. Regardless of the 'astronomical' odds in a lottery with umpteen millions of tickets, while it's highly improbable for anyone to win we know with certainty that someone will win. The minuscule chances of a win in a 6/49 lottery (approximately 1 in 14 million) doesn't prevent us from betting – and betting more when the pay-off is higher.

Behaving the way we do in the face of such probabilistic outcomes would have an observer think that we're risk-seekers. Yet, behavioural research has shown time and again that for the most part we're risk-averse. Overwhelmingly, we opt for a certain gain of $100 rather than a chance at a toss with an expected value of $120 – but which is subject to a 20% chance of gaining nothing, for example, $(80\% \times \$150) + (20\% \times \$0)$. More often than not, whether we trade off a higher expected value for a less-than-fair but certain outcome or choose instant gratification with less at the expense of having to wait longer for more, our relationship with probability deviates from the cold math and becomes subjective. At times, the notion of probability plays even more tricks in our minds. There's always a tiny probability for 17 consecutive occurrences of heads in coin tosses $(1/2)^{17}$. Following such a sequence, we somehow subjectively expect a higher chance for tails in the 18th draw. But the probability is still ½. This is the *gambler's fallacy*, equally applicable to casino games or the heightened expectation of having a son after five daughters. It takes a bit of mind-twisting but the central limit theorem is arrived at differently and doesn't justify such expectations.

In terms of applicability and usefulness in everyday life, the notion of probability appears better suited when statistical frequencies rather than binary events are involved. The probabilities derived from detailed underlying statistics make us seers in some sense.[‡] The restaurant owner has a good idea of how many patrons will visit his establishment on Mondays and Saturdays, and plans accordingly. We may not know who'll have an accident tomorrow but we have a good idea of how many will. The whole insurance business is based on such probabilities and sets premiums accordingly. When binary outcomes are at stake, such as rain or no rain, things get blurry. We express frustration for carrying an umbrella for nothing when the forecast showed an 80% chance of precipitation and dismay when we get soaked wet although there was only a 10% chance of rain. Such examples capture our unease with the notion of chance. They also demonstrate that the measures belong to an imaginary realm and only their shadows are reflected on our reality.

Digging a bit deeper, mathematicians realized that it's really the underlying *utilities* of the pay-offs or losses that matter more than the expected values. For the same cold mathematical probability, we react quite differently. We subjectively assign utilities to quantities of gains or losses – and not at all through straight lines. Behavioural research has shown that this depends on the *reference point*, that is, at what initial level we start. Our psychological reaction may also depend on time horizons and other preferences. What's more, there's an asymmetry in assigning subjective values to such utilities depending on whether a loss or a gain is concerned. Our risk-averse nature will settle for the certainty of a small gain rather than gamble for a much larger one with high odds but we'll risk away sizeable losses, with the same odds, to avoid a small but certain loss. These observations regarding asymmetry underpin the *prospect theory* of Kahneman and Tversky, which explains how and why the response to losses is stronger than the response to equivalent gains – something that doesn't follow the mathematics of probabilities. The clash between our rational and emotional sides is even affected by the framing (if not the wording) of the outcome choices. For instance, Kahneman provides the following example.

To fight the outbreak of a disease expected to kill 600 people, two alternative programs are proposed.

- Program A will save 200 people.
- Program B has a 1/3 probability of saving all 600 and 2/3 probability of saving no one.

---

[‡] This was the influence of Adolphe Quetelet who, over several decades in the 19th century, produced a bewildering number of detailed tables on every aspect of human life (including the body mass index). He was taken aback by the repeated regularity of crime year after year although, unlike regularities in the physical universe, it involves choice of action. Due to new norms, policies, and other actions, such 'social physics' (which eventually led to sociology) can and do change over time, though.

Although the expected value of B is still 200 saved lives ($600 \times 1/3 + 0 \times 2/3$), the majority of people prefer A. Saving 200 people with certainty appeals more than a chance of saving all 600 but risking a bigger chance of saving nobody.

In another version, though, the same two programs were framed differently.

- Under A' 400 people will die.
- Under Program B' there's a 1/3 probability that nobody will die and 2/3 probability that all 600 people will die.

In this formulation, the majority chose B'.

Where's the 'logic' in such behaviour? Since the consequences of the two sets of choices are mathematically identical, the authors conclude that when faced with choices between sure things and gambles, humans resolve them differently depending on whether the outcomes are good or bad. "… *the framing experiment reveals that risk-averse and risk-seeking preferences are not reality-bound. Preferences between the same objective outcomes reverse with different formulations*".[5]

In addition to utilities, reference points, and the asymmetry in losses and gains, the specific domain matters too. Money is one thing, life is another. Staring at certain death, a surgery with a 20% chance to live is much more appealing than a 20% chance to triple some considerable wealth or lose it all. Surely, such reactions test rather than refine our notion of rationality. In any event, our relationship with probability will continue to be work in progress due to our subjective interpretation of its meaning. Check with your neighbour, perhaps, if we've all come to terms that 2:1 odds and 66.6% probability are one and the same.

---

## Selected Topics

Data analysis is multifaceted. It extends to many areas and topics with their own peculiarities. Its primary objective is to explain our reality and its evolution. The quality of the questions asked and the interpretation of the data are among the crucial elements needed to succeed. The remainder of this chapter discusses a few select issues frequently encountered in socio-economic research, nationally and internationally.

### Stock-and-Flow Interpretation

It's not unusual for the analytical interpretation of data to stir controversy. Numbers don't have some built-in and widely understood meaning regardless of context. When a series is monitored over a long time, an analyst may be justified to say that during this period *only* 60% of the people participated in a certain activity – when up to that point the comparable number was hovering consistently between 80% and 90%. In other cases, especially one-off surveys, it's unclear whether a 60% engagement in something should be interpreted as high or low, and whether a 60%/40% split in support of a proposed action is justification enough to proceed. What would the difference be if the numbers were 70% for and 30% against?

The fact that we arrange our affairs and accounting in time periods, typically years, is not always in line with how we perceive the world around us. We understand stocks and flows and are familiar with the difference between deficits and debts or monthly household budgets and 25-year mortgages. We measure investment and depreciation as flows and understand how they impact the valuation of the capital stock. However, there are telling cases of protracted arguments over flow statistics because we're unaware of their stock-type lasting effects. Crime statistics is such an example. In recent years, many times and in multiple jurisdictions, annual statistics showed a marked decline in certain crimes. This doesn't sit well with many, it doesn't match what they 'experience' or their perception of crime in their societies. In such cases, questioning the completeness, the quality, and the very credibility of the crime data is logical – the alternative would be espousing massive paranoia. A 'third path' may offer a more likely scenario, though. The data are accurate and what's at play is a psychological effect invoked by the nature of the underlying events. While there's no reason why museum visits would leave any psychological trauma on the psyche of people, break-and-enters do. They're internalized and felt at a much more personal level, leaving scars on those affected for a very long time. Their presentation as annual flows doesn't quite fit the time they stay in people's heads.

The data behind the line in Figure 9.5 track plausibly the annual movement of break-and-enter statistics in a city of a million over a 30-year period. They were on the increase for a long 20-year period but they're on the decline in the last ten years. On average, 10,000 households have been subjected to such crime each year (1%). What this misses, though, is that the cumulative number of households affected continues to rise every year. Ignoring repeat victims, over the 30-year period from 1990 to 2020, there will be about 300,000 households traumatized by this particular crime, a substantial 30% of the population. Neighbours, extended families, and friends of the inflicted households are also exposed to their experience and the sense of being violated, adding to the stock of those feeling unease. Such thinking reconciles the lower annual statistics with people's sentiments.

Analogous may be the case for R&D statistics, which should not be expected to correlate to outcomes within the time period of their measurement. R&D projects may run for periods longer than a year and the cumulative spending on specific lines of inquiry would be more relevant for discoveries. (Recall

**FIGURE 9.5**
Stock and flow interpretations of data

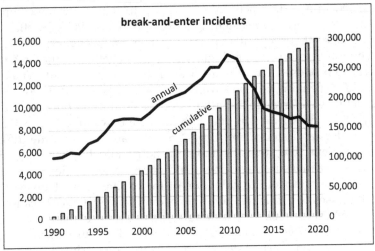

that R&D is now capitalized in the estimation of GDP.) Although such stock-type statistics are not currently produced, they may be helpful in the interpretation of phenomena along the lines discussed.

## Slippery Impacts

Actions such as government policies, business investments, or international aid aim at something. Whether solving a problem or seizing an opportunity, their ultimate target is an improved outcome rather than a tangible output. Filtering equipment for water wells improves community health through clean drinking water, a highway enables faster and safer transportation flows, a wireless tower enhances communication capacity. No one can fault the actors behind such actions for wanting to assess the impact of their interventions. In principle, they want to demonstrate success in achieving their objectives and making a difference.

Analytical work to quantify such impacts is akin to evaluation. It represents an application of the 'so what' questions that help both justify and refine work on ambitious ideas. It needs careful design, specific to the issue at hand, and appropriate measurement techniques and instruments. Ideally, data collection will be planned from early on rather than being an afterthought. No one can argue that the quest to understand impacts is misplaced. After all, in the final analysis, this is what really matters. Yet, for a number of reasons, including expectations for definitive answers, more often than not such work frustrates all those involved – sponsors, researchers, and receiving audiences alike.

A prominent example of impact research in the last two decades comes from the area of ICTs. It was understood from the beginning that the early emphasis on *access* to the Internet and other new technologies would soon give way to issues of *use* (frequency, intensity, type, and the like), which in turn will evoke the hierarchically more important question of *impacts*. Among early curiosities was the absence of expected economic impacts following the proliferation of personal computers. This was encapsulated in the *productivity paradox. ("We see computers everywhere except in the productivity statistics".)* Way too much energy was devoted to research that would eventually prove that the paradox wasn't a paradox at all. Simply, there were learning curves involved, needed reorganization of production processes, and time lags – not to mention the need for better measurements that were never 'above-suspicion' in that area. An alternative to that research suggested, in jest, to simply stand by and watch for businesses throwing out of their windows all their investment in computers and peripheral equipment if it wasn't of any use…well, that didn't happen.

The question of impacts was better framed later when households, businesses, and governments adopted massively and rapidly a large number of technologies, gadgets, and applications. The Internet, websites, e-commerce, broadband connectivity, laptops, and smartphones all exerted independent influences. At first, a quick, if sneaky, attempt to peak at impacts was through the measurement instruments for access and use – predominantly surveys. After asking households and businesses a barrage of questions on their adoption and use of ICTs, why not stick a couple of 'impact' questions at the back? Do you think that your life is better now, your business more productive? But when someone goes on record with how much they got, how much they pay, and how much they use, what answer is supposed to follow? Would there be any good use for answers like 'eye-opening' and 'awesome' or 'nothing' and 'waste of time and money'? Surveys were pushed farther than they should, with disingenuous and leading questions inviting self-serving answers commensurate to the expenses and the efforts invested on ICTs by the respondents.

This was neither an ideal nor a meaningful way to get to impacts. To study poverty, we don't ask people *"are you poor?"* but we collect income information and draw poverty lines as we see fit. Inferential analysis is needed, based on solid data. To unravel the impact of ICTs on firm performance, survey data must be linked with measures of productivity, profitability, market shares and the like extracted from independent sources, such as income statements. Thankfully, in time, such studies were performed. Similarly, rather than self-assessment of impacts, ICT survey data for families and individuals can be linked with independent data on time use, relationships with family and kids, friends and neighbours, and a plethora of other things. The coexistence of both positive and negative impacts must be addressed explicitly.

Generally, research on impacts must cope with two fundamental issues. The first relates to the type of the impacts sought, whether economic or social, applicable to the community at large or specific subpopulations. Often times both quantitative and qualitative data collection methods are used, such as standardized surveys and open-ended 'key informant' interviews that may lead to contradictions. The second issue relates to the time horizon. How long after an intervention or an event do we start looking for impacts, and for how long are they supposed to last? For both issues, it's imperative that the scope of the exercise is well defined, and pragmatic expectations are set through appropriate and clear choices. At times, however, sponsors of such studies show ambivalence in distinguishing between communal and individual well-being or reluctance to settle for benefits of a certain duration only.

Consider the following real example. A donor organization invested in the building and operation of a telecentre in a remote village of a poor country, when no computers or Internet access existed. In due time, a study on impacts revealed a substantial return-on-investment. Hundreds of people from the village had become computer and Internet users, many regularly, some occasionally. They learned how to use the new technologies, got valuable information for their agricultural activities, learned more about their health, education, and foreign affairs. Many youth cut their teeth on the new technologies there, something that will forever mark their lives, and even started to teach others. What's more, it looked as if the financing model used could become self-sustainable, with small fees. Quite a success story. In parallel, though, a wandering researcher conducted informal interviews in the village and vocalized a contrarian view. The telecentre as it was run harmed life in the village. The reason? Not that only few people used it, that it wasn't affordable, or it didn't open new opportunities for many. The fellow who was chosen as the operator had a feud with a cousin. Members from the families siding with the cousin would never step in there. Under the circumstances, the telecentre accentuated a divide harmful to the social cohesion of the village. (The researcher also got a statement from a user at the telecentre that her life will never be the same.)

Through a similar initiative in another country, someone got an education in accounting and started doing good business there and in surrounding villages. Soon enough, though, his good fortune attracted the animosity of others to the point that he and his family could no longer live there. Not the economic benefit and wealth-creation through upskilling that the donors had in mind for their investment. The morale is that the chain reaction of events has no end. It can lead to labyrinthine corridors, with endless anecdotes to navigate. In the real world, it's tough to achieve the economists' *Pareto optimal*, improving societal welfare by making at least one person better off while not making anyone worse off. Human nature rubs off against all kinds of impacts – even jealousy.

Impact research cannot be forced to levels impossible to defend. It's unintelligent, for instance, to look for the impact on the GDP from the opening of two telecentres somewhere. An additional reality check comes from internalizing the mere fact that our world is dynamic. Both donors and government bureaucracies struggle to come to terms with fleeting and ephemeral rather than eternal success. However, progress and evolution are subject to creative destruction processes. It's clearly not sustainable to keep piling the new on top of the old indefinitely. What outlives its usefulness must give way. Simply put, project success cannot be defined as eternal life but making a difference during its life.

An apt example comes from the *Grameen phone ladies*, a project that became the poster child in the development arena in the early 2000s.[6] The Grameen Bank, created decades earlier by Muhammad Yunus to offer low-cost microcredit to those with no collateral and for which he was awarded the Nobel Peace Prize in 2006, enabled women in rural Bangladesh to set up mobile phone exchanges. For a period, these micro-businesses did a lot of good, directly and indirectly, by helping many communicate at very low prices, boosting entrepreneurship and alleviating poverty. That the scheme didn't last, in light of all subsequent advances, doesn't subtract anything from its impact at the time. Rather than something in the dustbin of history today, its true contribution must be examined against its own time. Mindsets can come around to that contributing a step in a staircase is indeed a worthy contribution.

The roller coaster of impacts vis-à-vis the time trajectory is well captured by the Taoist tale below, which is recommended for all those who have absolute views and demand 'clear and conclusive' answers.

*An old farmer works his crops for years. One day his horse ran away, quite a loss. The neighbours said sympathetically "What a shame, such bad luck"! The farmer stoically replied "Maybe". In a couple of days the horse returned and brought along three more wild horses. "What an amazing turn of events, such good luck" exclaimed the neighbours, to which the farmer again responded "Maybe". The next day, trying to tame one of the wild horses, his son was thrown down and broke his leg. The sympathetic neighbours again expressed their sorrow for his misfortune, eliciting another "Maybe" from the old farmer. When a few days later military officers came to the village to draft young men for war, the son was spared. The neighbours congratulated the farmer on his good fortune. They got back another "Maybe".*

## League Tables

This moniker applies to ranked lists of the type used in sports and elsewhere. League tables are also commonly used for inter-country comparisons in a thematic area. They represent analytical activities in that they involve the conceptualization of a 'framework' and the manipulation of data. Their construction utilizes a variety of methodologies, which always include *aggregation* across indicators of interest primarily in the form of composite indices. This follows the prior *normalization* of indicators, conversions of the data used as proxies to the same scale. Sometimes subjective weights are also applied. The availability of data dictates the number of countries included in individual league tables, which may range from a few tens to nearly 200. Invariably, such exercises run into data gaps, which are dealt with through imputations or – if financial resources permit – some quick-and-dirty surveys soliciting 'key informant' perceptions. More timid cultures are more modest in their responses. Frequently, the same data are used to proxy different indicators, for example, the GDP is used for income or the standard of living. This partially explains why some league tables in different topics look very much alike – with the same culprits on the top and the bottom.

A plethora of league tables exists for any number of topics. Most are short-lived and little known beyond their immediate circles. Some are well known and have become media darlings, attracting significant attention. This is the case of the UNDP's HDI, and several others colloquially known as Quality of Life indexes. The HDI aggregates across gross national income per capita, life expectancy at birth, and education, based on years of schooling.[§] The HDI is then calculated as the geometric mean of the normalized indices for the three dimensions.[7]

More often than not league tables are intended as advocacy instruments to draw attention to a subject matter by attracting popular press rather than reveal deep analytical findings. In that sense, they also help popularize the importance of statistics among the general public. Yet, at times they consume more energy than warranted at the expense of other analytical activities. Having professionals in conferences seriously asking why their country is 45th in one index but 57th in a somewhat similar exercise is not the pinnacle of international brainstorming – particularly when it's immediately obvious that one exercise includes 90 countries and the other 135. Generally, the biggest flaw of such indices is that their 'frameworks' are not cohesive conceptual constructs. As a result, they don't measure the evolution of a notion before they focus on comparisons across countries. Some claim they do but when the ranking of countries changes from year to year we can only know that a given country fell two spots behind, not how it fared against itself. For that we need to resort to the individual indicators, one at a time. There are exceptions, involving the conceptualization of notions that can be measured independently and over time. One such example was the Infostate model, used for the measurement of the digital divide internationally around the twin Worlds Summits on the Information Society (2003 and 2005).[8] Such notions are conducive to measurements similar to that of the GDP. As a particular notion of output, it can be seen as an aggregate and composite index, which allows a country to monitor its own performance over time and enables comparisons across countries.

---

[§]   The indicators are converted to indices with values between 0 and 1 using minimum and maximum values (goalposts), as: Dimension index = (actual value – minimum value)/(maximum value – minimum value).

## *Fact-Checking Tips*

*Seeing is not believing:* Economic priorities take turns. At times, inflation is the big enemy, other times unemployment or the budget deficit are the evils to fight against. In the aftermath of the global financial crisis, the pendulum had swung on the need to tame debts. The same fiscal and monetary tightness that almost killed the euro were advocated by many on both sides of the Atlantic. A 2010 paper, "Growth in a Time of Debt", provided ammunition to the austerity cause, particularly since it was penned by two Harvard economists, Carmen Reinhart and Kenneth Rogoff.[9] The key finding from a large study of 20 countries in the post-war period was that when a country's debt-to-GDP ratio hits or exceeds 90%, average real economic growth slows – albeit only by 0.1% it was still an arrow in the quiver of austerity proponents.

This finding was used and reused repeatedly. There was no reaction until students in a graduate course at the University of Amherst had to replicate the results of a paper. Thomas Herndon was struggling to do so for this particular one. While initially calculation errors by the likes of the particular authors didn't cross his mind, his fact-checking came to the point to contact them for the data – which came together with calculations in Excel spreadsheets. He then spotted that in calculating average GDP growth the authors hadn't selected the entire row, missing 5 of the 20 countries – Australia, Austria, Belgium, Canada, and Denmark. Working with the professors, Michael Ash and Robert Pollin, they discovered more questionable issues. There were missing data, and some funny weights. A paper they published in April 2013 made the headlines. In their own words, in replicating the study, they found that "...*coding errors, selective exclusion of available data, and unconventional weighting of summary statistics lead to serious errors that inaccurately represent the relationship between public debt and GDP growth among 20 advanced economies in the post-war period. Our finding is that when properly calculated, the average real GDP growth rate for countries carrying a public-debt-to-GDP ratio of over 90 percent is actually 2.2 percent, not –0.1 percent as published in Reinhart and Rogoff.*"[10]

This reversal in average GDP growth, from –0.1% to +2.2%, was a big and high-profile affair. There are many more 'Excel errors' that don't receive the same publicity. Such stories not only boost the case for replication but also reveal the immense usefulness of fact-checking. *"Measure twice and cut once"* is always good advice. While the implication is not that every fact-checker should obtain all the data and all the calculations of every paper or report, this story highlights shortcomings in existing processes, including peer reviews, and points to the importance of alternative verification processes. It also demonstrates that as much as there's glory in analytical findings, there's glory in fact-checking – Thomas Herndon can attest to this. The asymmetry that there's no glory in fact-checking if no errors are found will always be there, but then not all analyses result in glorious findings either.

*Seasonality:* We know that sub-annual data can be noisy. Part of the characteristic ups and downs of many time series are rather predictable and are explained by *seasonal* and *calendar* effects. Retail sales are higher in December than in January, February has only 28 days, some months have five weekends and the like. Frequently, such effects are removed from the data through *seasonal adjustment* techniques. The idea is to put sub-annual data, typically monthly or quarterly, on an equal footing in a way that their *trend* is easily visible. Seasonality can obscure the true, long-term movements of a series, such as the peaks and troughs of a business cycle. Stripped of changes that occur at the same time and approximately with the same magnitude, seasonally adjusted data are considered more meaningful for the interpretation of economic conditions over time.

Familiarity with seasonally adjusted data is very material to fact-checking. First, differences between raw and seasonally adjusted estimates can be so sizeable as to reverse even the direction of the change between periods. As a fact-checker you need to know what data you're looking at and avoid apples-to-oranges comparisons.

Second, you need to know which data are more appropriate for certain uses. Raw data are preferable for series with no identifiable seasonal pattern. Ditto for raw data that represent the true metrics of interest, such as monthly movements in CPI prices or the number of employees in an industry. Seasonally adjusted data are more conducive to longer-term investigations of trends although, for really long periods, why use sub-annual data at all becomes a relevant question. You'll also encounter year-over-year comparisons of raw data, as an alternative to seasonally adjusted data. Since the same month is compared over two or more years, an argument can be made that this neutralizes seasonality – but the danger comes from unusual months and years. While seasonal adjustments end up dampening the turbulence in sub-annual data, other smoothing techniques do so explicitly, such as *moving averages* of various durations. With such methods even more different data are now circulating.

Third, you must be aware that even after the removal of seasonal and calendar effects from a time series, the remainder is not only the trend but includes an *irregular component* too, which accounts for events like strikes, natural disasters, or other 'noise', including sampling and non-sampling errors in survey data. It may not be readily discernible if some of the movements in the data represent short-term and transitory fluctuations or are indicative of structural change that marks a sudden but permanent deviation from the past.

*Handle with caution:* (i) We discussed fake correlations earlier in this chapter. Some can be funny and, if you have the time to invest, you can get a kick out of them. Then, there are others that pass as serious findings because of flawed analysis. In 2015, a story that made the rounds was centred around a chart showing that musicians involved in newer genres die young. Tracing the numbers back to work by the original author,[11] the chart looked like the

**FIGURE 9.6**

Life expectancies of male musicians by genre

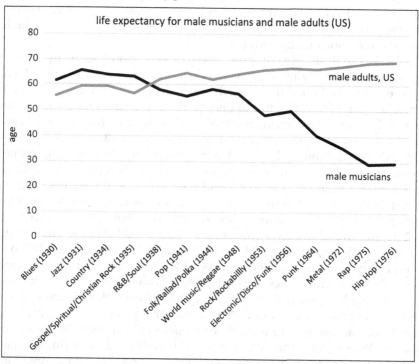

one reproduced as Figure 9.6.** The average year of birth of male musicians associated with a given genre was estimated, as was the average age of death of those who died. This was then plotted against the life expectancy of US males during those years, and shows dramatically shorter life spans for newer genres. There had been several shootings and suicides leading to premature deaths of artists in the newer music genres. Still, the numbers looked astonishing. Average age at death for rap and hip hop dropped to under 30!

Now, there are several issues with the data in this story. It'd be worth looking at the microdata behind the average year of birth of the musicians included in the different genres of music, as well as the average age of their deaths. Even overlooking that, Carl Bergstrom and Jevin West in *Calling Bullshit* explain the flawed finding as a case of truncated series.[12] For the newer music genres,

---

** The Washington Post article "Why a musician's life expectancy depends on what kind of music she plays" was reproduced through cut and paste. To make the effect more dramatic, the trick of truncating the vertical axis was also used – it started at 25 rather than zero, making the drop visually more precipitous. The chart also separated male and female musicians. Figure 9.6 is based on male musicans and male life expectancy.

the numbers include only those who had died – who are also the ones who died prematurely. All those alive at the end of the period were excluded. The life spans of dead taxi drivers born in the same years would have been cut short too. We'd discussed earlier that truncation, on the left or the right, is a key reason for such faulty correlations.

(ii) In data analysis, *rounding* numbers is common. You'll see many footnotes in statistical tables saying that *"numbers may not add up due to rounding"*. With a few exceptions, I don't like more than one decimal, and I frequently prefer whole integers. However, this applies to the presentation and communication of final numbers. During the analytical work itself (the 'spreadsheets'), numbers used as inputs in calculations must be maintained in their most detailed form. Rounding 1/3 to 0.3 at the processing stages will compromise the accuracy of the final numbers, even more so if repeated rounding is practiced, for example, 1.45 to 1.5 and then to 2. You'll meet discrepancies in numbers that are due to such practices. A classic example is that if you multiply 1.5 by 1.5 the rounded answer will be 2 but if the 1.5 is rounded to 2 beforehand, the answer would be 4 – and would stay the same for any number up to 2.49, although $2.4 \times 2.4 = 5.76$, which would be rounded to 6.

(iii) Lastly, visualize a chained gate or a locked door in a service room with a big skull and crossbones sign explicitly warning *"Danger: Do not enter"*. Most people heed such warnings, we move on even though we may suspect it's an exaggeration. I can't help letting this picture in my mind every time I come across one of my all-time favourite data aphorisms: *"handle with caution"*. What is someone supposed to do? What's the data world equivalent for hazmat suits? Are rubber gloves and lysol needed before data contact and use? Such warnings are offered by data producers for low-quality data with big standard errors or similar measures. This is particularly the case in survey data, where the quality thresholds targeted for the key variables deteriorate with detailed cross-tabs from thinner cells. The range within which mean estimates lie gets quite stretched. Past a certain point, the source may not even publish such estimates – which may still find their way out as there's no legal requirement to the contrary. With no real confidence in the estimates, the monkey is passed to the user. The implication for fact-checking is that the quality status of the data is also needed. If the main conclusion of a story you're checking hangs crucially on a 7, not an 8, and the 7 could be anywhere between 3 and 11, you need to call it out.

## Notes

1  This organization works on map data and describes itself as: "OpenStreetMap is built by a community of mappers that contribute and maintain data about roads, trails, cafés, railway stations, and much more, all over the world." © OpenStreetMap contributors, www.openstreetmap.org/about

2 For instance see, Tyler Vigen, "Spurious Correlations," *Tylervigan.com*, http://tyl ervigen.com/spurious-correlations

3 To which mathematician Ellenberg says yes, and uses it to provide proof that 0.99999 = 1.

Jordan Ellenberg, *How Not to be Wrong: The Power of Mathematical Thinking*. New York: Penguin Books, 2015.

4 This was explicitly discussed by Nassim Nicholas Taleb, *The Black Swan: The Impact of the Highly Improbable*. New York: Random House and Penguin Books, 2007. For related matters, see also Leonard Mlodinow, *The Drunkard's Walk: How Randomness Rules Our Lives*. Toronto: Vintage Books, 2009.

5 Daniel Kahneman, *Thinking, Fast and Slow*. Toronto: Anchor Canada, 2013, 367–368.

6 See, for instance, USHA, "GrameenPhone's Success, Phone Ladies' Loss," *Appropriate IT*, September 17, 2007, www.appropriateit.org/grameen-phone-lad ies/

7 United Nations Development Program (UNDP), "Human Development Reports: HDR Technical Notes" https://hdr.undp.org/en/content/hdr-technical-notes

8 George Sciadas (Ed.), *Monitoring the Digital Divide...and Beyond*. Montreal: Orbicom), 2003, https://orbicom.ca/wp-content/uploads/2017/06/Monitor ing-the-digital-divide.pdf; George Sciadas (Ed.), *From the Digital Divide to Digital Opportunities: Measuring Infostates for Development* Montreal: Orbicom, 2005, https://orbicom.ca/wp-content/uploads/2017/06/From-the-Digital-Divide-to-Digital-Opportunities.pdf

9 Carmen M. Reinhart and Kenneth S. Rogoff, "Growth in a Time of Debt," *American Economic Review, American Economic Association* 100, 2 (2010), pp. 573–78, 2010. Retrieved at: https://dash.harvard.edu/bitstream/handle/1/11129154/Reinhart_Rogoff_Growth_in_a_Time_of_Debt_2010.pdf?sequence=1

10 Thomas Herndon, Michael Ash and Robert Pollin, "Does high public debt consistently stifle economic growth? A critique of Reinhart and Rogoff," *Political Economy Research Institute (PERI), University of Massachusetts Amherst* (Working Paper Series, No. 322), April 2013. Retrieved at https://peri.umass.edu/images/WP322.pdf. Also published later in Cambridge Journal of Economics 38, 2 (March 2014), pp. 257–279. Retrieved at: www.jstor.org/stable/24694929

11 Dianna Theadora Kenny, "Why a musician's life expectancy depends on what kind of music she plays," *The Washington Post*, March 31, 2015, www.washing tonpost.com/posteverything/wp/2015/03/31/why-musicians-life-expectancy-depends-on-what-kind-of-music-they-play/; See also PP Presentation by Dianna T. Kenny, "Music to die for: Pop musicians are at increased risk of early mortality and morbidity," *ResearchGate*, July 2015, www.researchgate.net/publicat ion/320567595_Music_to_diem_for_Pop_musicians_are_at_increased_risk_of_early_mortality_and_morbidity

12 Carl T. Bergstrom and Jevin D. West, *Calling Bullshit: The Art of Skepticism in a Data-Driven World*. New York: Random House, 2020.

# 10

## *The Future of Data*

With the first two decades of the 21st century behind us, we're currently in the vicinity of the end-of-the-beginning of what is referred to as a *data revolution*. While the particular phraseology we choose to describe what we're experiencing will not determine the way forward, revolution implies throwing out an old order and replacing it with a new. We find ourselves in the middle of much commotion, hype, lofty promises, and a fair amount of confusion. Revolutions are not known for their orderly steps after all. Perhaps, a more apt interpretation of what's happening is a series of tectonic shifts, with temporary rearrangements of pieces that last only until the next aftershock. We need to brace ourselves since multiple collisions among different paradigms are happening simultaneously. In the process they force open an array of issues, from data ownership to the privacy implications of their reuse, and everything in between. Our own responses to the frictions caused by such issues will determine the future.

There's no future path for data over the next few decades independent of our responses, which are far from clear. The sure thing is that the old order, as we knew it, is done. Instinctively thinking of the statistical office when it comes to data will be no longer. But it'd be premature to make any pronouncements of death of old-order players or assess the vim and the longevity of new-born ones. A perceptive distinction between pivotal forces and individual players is indispensable to understand the new landscape, and firmer decisions will be necessary to cut through the many unknown paths created by the tremors before alternative epicentres emerge. However, we've already entered a period of both introspection and forward-looking thinking on how to plough ahead. We'll be better equipped if we understand the higher meaning of forces at work, and where they fit in the big scheme of things. This is especially so if, individually, we're to contribute to our collective responses. Familiarity with the modus operandi of how data are produced, how they flow, and how they're used is essential.

The future of data will be complex. It'll contain bright areas and dark alleys. Looking at my crystal ball I see that:

- the future will be awash with data, much more of what we know now and then some
- data production will be mainstreamed, happening in multiple public and private organizations

DOI: 10.1201/9781003330806-10

- players will be frenemies, engaging in short-term collaboration and long-term competition
- statistical products impossible yesterday will become commonplace, whereas the production of relatively simple statistics today will become more cumbersome
- algorithms and processing code will be the new currency, transforming raw data into usable statistical products, and doing so at multiple layers
- statistical conferences will require big-tent venues hosting many new specialists, including a yet-to-be-named breed of a methodologist programmer
- governments will have strong opinions on matters of access, and 'trusted third parties' will emerge as key players in the new data ecosystems
- research will produce more data-driven, theory-free findings, subject matters will get mixed up with methodological and even technical matters
- data boutiques and resistance movements will spring up
- 'truths' will become more elusive, research results will be questioned more
- we'll be talking more about privacy – and live with less
- numeracy efforts will multiply, as will gaslighting
- machines will take over, automation in data production will be followed by user machines

Just kidding, though. I'm not a futurologist, and I don't have a crystal ball. But one is not needed. All these pronouncements are happening now. As will be explained below, they offer important clues on where things may be headed.

We can't exhaust all possible issues related to the future of data in a closing chapter, but we can certainly discuss some big ones. We're not looking for a unique answer. There's no karma leading to a final destination. Our objective is more like positioning the big landmasses on a map, and pinpointing our own coordinates. Then we can plot paths for any destination we desire. Since our own responses to some of the key issues will determine the future, best we can do is look at our likely responses. For that, we can employ all we've got. History, experience, intuition, foresight and data. Other than that, there'll always be room for betting men when it comes to the endless 'details'. Where will the data reside? How will they be processed, and by whom? Will researchers have access to confidential data? Will microdata linkages among unlikely sets actually happen? Are there opportunities for new players? What will governments do? Do we still need models? We'll deal with as many such questions as we encounter in our path.

## A Messy Neighbourhood

A good first step is to solidify our understanding of where things stand today. A bird's eye view of the new data world could look like a funnel, where multiple data sources are siphoned in through its wide opening and multiple data outputs exit on the other end. The same source can support multiple outputs and a single output may require data from multiple sources. Everything that matters happens inside that funnel. This high-level view, however, would conceal important relationships among data sources, and will totally miss the data hierarchies involved. Placing the emphasis on the production, flows, transformations, and uses of data will better prepare us to identify the relevant questions ahead and explore our choices.

An abundance of data is generated by traditional and new sources. In a dynamic world, some sources will intensify their data throughput, others will dry up, and new sources will be added. Our conceptualization shows data sources as dots in Figure 10.1. But this abstraction should not fool us. Each dot contains a lot of raw data. They could be the completed questionnaires from a survey or the census forms containing data on individuals and their families. They could be the debit card transactions of a bank's customers, the scanner data of a retailer, or the call records of a mobile phone company. Google searches, social media postings, or tweets can be other dots. Some dots are bigger than others (not shown to scale). Facebook will accumulate more data in a day than a retail outlet ever will.

At their most granular level, data contents are heterogeneous. Some were collected by design and significant upfront effort, and have undergone sufficient processing that's made them ready for use. Traditional microdata would fall into this category. Others grew organically, they're fragmented, scattered around untreated, and not ready for primetime. This is so because most of the new data were generated without any thought to their eventual statistical uses. The motives behind a single purchase, a telephone call, or a tweet had nothing to do with statistics. Businesses collect them as parts of their daily operations but statistical uses aren't top-of-mind when engaging in such routine activities. They definitely don't enter the mind of sensors programmed to send signals, generating more of these dots. In that sense, while the data contents inside the dots may be large, their information value is minuscule or next to nothing. To become fit for statistical action they need some tender loving care.

As we saw in earlier chapters, extracting information from data requires treatment through processing, aggregation driven by the desired outputs, and the like. This is more so for the new data whose constituent components are at the transaction level, below the traditional micro units. Moving from data smithereens to usable microdata requires consolidation to relevant units, such as individuals or households, and conversions to useful time periods – among others. Such 'specs' will pass data points a, b, and c through a 'black box' and transform them to usable data sets, the white circles A, B, and C.[*] For a long time now, the contents of such boxes come largely in the form of computer code. While the specialized nature of coding posed various challenges inside data producers, it's increasingly mainstreamed and more people are versed both to write and decipher code. In fact, unlike the underlying confidential microdata, the code can be made public. This is particularly the case of open data, where storage and reuse of code are actively encouraged.

Things start to get interesting early because decisions already need to be made to arrive at this higher layer of data. Methodological choices will be subsumed in the code, in a way that the intended use of the transformed data comes into play. Clearly, this is not a unique exercise. For instance, aggregating daily totals from barcode scanners for a retailer to know his Monday sales may only require a simple addition. If unique customers were the target, identified perhaps through debit cards, another approach would be needed. The same customer could have made back-to-back purchases a few minutes apart. If the data are needed to replace survey data for the CPI, the specs will be different still, to match the timing and products of interest and avoid the noise of individual transactions. In most cases, the existing granularity in the raw data can support many different data transformations. It all depends on how much 'reality' we want to uncover for our intended uses.

Different transformations from the same data source are possible, such as D1 and D2 from source d, either through an expanded black box or additional ones. As well, usable data sets can be produced by combining data sources through linkages, such as set E from sources e and f. However, there are time and cost implications associated with any black box and there's an incentive at large for data sets at that higher layer to serve as many purposes as possible. Investments dictated by a primary use may or may not be conducive to alternative uses. Either way, this is still one step and not the end of the production process. Additional data sets can be created across data sources at higher layers through analogous processes, involving more specs and black boxes. In our conceptualization, the highest levels are shown as stars. In Figure 10.1, star R1 culminates from data sets A, B, and C, and star R2 directly from data sources e and data sets E and F. Such stars already exist in the statistical system through decades of evolution and investments. They're the modern statistical registers in the sense we discussed in Chapter 8, and which are now capable of linking collections of microdata from multiple sources. The code is already there and is expanding as appropriate. This

---

[*] A black box typically signifies that its contents are a mystery to the observer – but only until it's opened. Then, like in airplane flights, it contains all the information needed. In that sense, you can simply substitute 'box' for 'black box'.

**FIGURE 10.1**
Data transformations and use

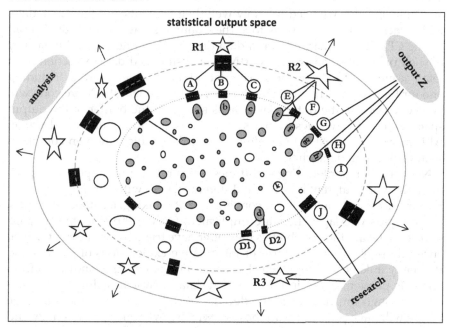

layer also defines the destination paths for the ongoing generation of data. All the new birth registrations, for instance, will start as a dot but, through known black boxes and pathways, will end up quickly as one more year in one variable of a population register. All new business created will soon end up as one more addition in the same variable in the business register. Much of the traffic jam and potential mishaps in the chaotic landscape of Figure 10.1 will be avoided.

Through similar investments, including trial and error, some of the new data will also evolve to reach a similar level of maturity and take their rightful place in the constellation layer. Assume, for instance, that A, B, and C were created individually by credit card companies for their respective transactions. Since each of them only has a piece of the total, they may decide (or be asked by the government or a 'trusted third party') to follow all the necessary steps and take it to the next level, aggregating across the whole country. R1 would then become a powerful resource, looking very much like a statistical register due to its census-like coverage. This would be easier if A, B, and C are comparable in their design and make-up. The same can be true of retailers, mobile phone companies, or other like businesses. Efforts like these will keep many people busy for years to come, but will create new stars!

Now, the production of statistical outputs or research and analytical activities can proceed. We can choose liberally among any dot, white circle, or star available. For this we need to know well their contents and their metadata, which include the black boxes too. The example shown as Output Z could be the CPI. Rather than rely on a data dot, representing a monthly survey of shelf prices, it's constructed partly by a customized data extraction from retail scanners (G), partly from web data (H) and other sources, including one for the weights (I) and a smaller survey component (h). More code will be needed before the output level, as is shown at the border of the output space. Alternatively, a researcher's project design can tap dot k and use an API (black box) to extract Google data (star R3), which are then combined with geolocation data (circle J).

Now, all that is interesting – and messy. Everything is happening experimentally, with different motives, at different speeds, with varying degrees of rigour, and we're nowhere near convergence of thinking, let alone standards. Some efforts overlap, others are duplicated. On the flip side, the proliferation of data and the possibilities they offer have triggered excitement, which currently plays out in the form of imaginative explorations of sources and the many linkages among them. The separation of genuine innovation from anarchic pandemonium in this crowded data ecosystem will happen gradually through much work in the decades ahead. The sure thing is that the future will have plenty of black boxes – which now house methodology and code jointly, including peculiarities of software. Moreover, precedents set today will affect newcomers tomorrow.

Inevitably, we'll experience the embarrassment of riches, in the sense that multiple outputs with conflicting messages will be proliferating. This may also lead to some disrepute and *statistical lawlessness*, something hard to get used to. A flavour of that was tasted in a plethora of studies which, in quick succession, used mobile phone data to track detailed population movements during the COVID-19 lockdowns. They were definitely relevant under conditions of a pandemic but their policy impact was constrained because, as a whole, they weren't deemed reliable. Both the particular data sets employed and the black boxes used were criticized. Non-transparent metadata and inconsistent messages didn't help either. However, not only they represent a useful experience-building step in the process but added to the body of evidence that researchers can access and use such data.

## Co-habitation and Strange Bedfellows

The value of data is in their use. And, thanks to the inquisitive nature of humans, they'll always find a way to be used. To start with, they're used by their sources for several purposes. Some to run and improve their business models, others have already started using them for research and statistical outputs too. Some do so on their own and opportunistically but, as time goes by, some others may well be dragged into it – legally or morally. Such developments suggest that in the gentrified data neighbourhood the statistical office won't be the first or only outfit that jumps to mind.

A bigger point is that data plays are made by data players. The crowded new ecosystem has led to confusing lines between research interests and possible data sources on offer, as we saw in the example of measuring a few backyards in a neighbourhood in Chapter 2. Menu costs are now higher, coupled with a murky understanding of what's permanent and what's transitory and passing by. Old comfort zones are lost and new domains to explore have appeared in areas not previously visited. Older and newer neighbours feel each other out. They're all under pressure – of a different kind. Some for survival, others for relevance, some for the opportunity to shine and show off, others may not quite know why – they just enjoy the ride or got caught in the crossfire.

However, the action has expanded and many want a piece of it even when the rules of engagement and guidelines are still absent. We've already argued that knowledge of the mere existence of data incentivizes demand. While the modalities of obtaining the data or gaining access to them is far from established, the cat-and-mouse game is always on. Alliances that confer short-term gains may or may not last the test of time. We've already seen that researchers manage to get to confidential data through various means, including universities, local authorities, and private businesses. Many statistical offices have had protocols for microdata access for some time. Social media and others offer access to parts of their data holdings through APIs, as well as invite researchers in-house. Financial institutions have started doing the same. MasterCard is involved in *data philanthropy* through research partnerships. Facebook's *data-for-good* program got involved in COVID-19 efforts. *Data pooling* has become 'a thing', where data are shared together with expertise and resources. In the absence of a playbook, which has yet to be written, data alone appear to be driving such efforts. Other forces will join soon, though. In the meantime, jockeying for position in this new environment is creating circumstances that until recently would have been considered unthinkable or weird.

Mark Zuckerberg was a student in Boston when Facebook was launched. He wanted to connect friends. Fast forward a mere 14 years, and try to imagine if this 2020 release could have appeared in his wildest dreams those days: *"The Future of Business Survey is a new source of information on small and medium-sized enterprises (SMEs). Launched in February 2016, the survey – a collaboration between Facebook, the OECD, and the World Bank – provides timely information on how firms with a digital presence assess the current state and future outlook of their business, the main challenges they face and their involvement in international trade. In 2020, from May to October, the Future of Business Survey was conducted on a monthly basis to take the pulse of how SMEs are navigating the*

*COVID-19 pandemic. The December 2020 survey will see the Future of Business Survey return to its biannual cycle.*"[1] The OECD participates in the design of the survey, which covers SMEs with Facebook accounts across more than 40 developed and emerging economies, and touts it as "*an innovative experiment of public-private partnership in data development and collection*".[2] Facebook's involvement is an aside to a busy agenda, which includes influencing policies by the Congress, banning a former US president from the platform, and building a metaverse.

Just in case you think this is an aberration, here are parallel 2019 releases from Microsoft and Statistics Canada.

"*Following our release of US buildings footprints last year, we've been looking for new markets to apply our techniques, and opportunities to continue our commitment to the open data community. As a result, the Bing Maps Team collaborated with Statistics Canada to deliver these 12 million building footprints, released as Open Data.*"[3]

"*The open microdata used to create version 2.0 of the ODB came from 65 different data providers from January 2018 to February 2019. A complementary database, constructed by Microsoft, using deep neural networks to extract building footprints from satellite imagery is also available on the Microsoft GitHub Repository.*"[4]

Telefónica signed an agreement with the UN's Food and Agriculture Organisation in 2017 for data-related initiatives related to forced migration in Colombia. Vodafone touted a 'pioneering programme' to help governments track population movements and control epidemics through aggregated and anonymized data. Twitter collaborated with UNICEF to track anti-vaccination sentiments in eastern Europe. In 2018, Statistics Netherlands entered an agreement with MarineTraffic to share vessel tracking data. Many statistical offices have already made deals with retailers, cell phone companies, utilities, and other data sources, and they're either at advanced stages of experimentation or have actually started using scanner data for the CPI and other products. Such data are expected to improve the quality of existing outputs, enable the production of new outputs, and reduce costs. Additional impetus on all such efforts came under the conditions of the COVID-19 pandemic.

These few examples barely begin to scratch the surface of the flurry of ongoing activities. Such happenings are behind the view that the existence and utility of such data is a *force majeure* for future developments. There is no stopping such force, notwithstanding matters of perceived ownership or confidentiality concerns. The likely response to such challenges will be the intensification of efforts to solve or manage them rather than giving up on prospecting for gold. We can't expect less deal-making in the future. The incentive structure is all there, from 'doing the right thing' to being socially responsible. None of that needs to be doubted, yet not all of the wheeling-and-dealing underway will have lasting power – some will be of the flavour-of-the-month variety. All in all, I fully expect much more use of the data, not less.

Indirectly, such discussion starts to answer the big question *"whose data is it?"* The involvement of governments in all that cannot be underestimated, and will help the arrival at a more complete answer.

## Data as a Strategic Resource

In the geopolitics of our times, data are exalted as an essential resource that will not only stimulate growth and prosperity but will determine the future balance of powers. An artificial intelligence race is on, contingent on data. Any forward-looking vision calls for data to be used and reused, again and again. This is what adds value and creates wealth. The use of data will expand existing businesses and create new ones. It'll also improve government. For society, there are gains all around. Such pronouncements and expectations hold clues for what lies ahead but, to materialize, data must be accessible. Solutions to overcome hurdles will be sought, of course – but the data will be used.

The EU has been particularly assertive, putting forward less-than-subtle answers. Pointing to the wealth that can be squeezed out of existing and forthcoming zettabytes of data, the Forward to a 2020 report by an expert group set up by the European Commission states: *"This creates an extraordinary opportunity for Europe to use this enormous amount of data yet to be created and lead the data revolution on the world stage."*[5] In addition to data-hungry AI applications and the transformative impacts expected from business-to-business data sharing, improvements in the delivery of public services that can come about from government use of private data are singled out. The tone is on *how* governments can help it happen, not *if*. *"… most of this much-prized data is in the hands of businesses, not of public authorities, with the latter lagging behind in embracing the power of data to inform their daily policies and service-delivery actions. Hence the challenge…to explore the creation of an enabling environment for privately held data to be shared with (or at least be accessible to) public authorities in complying with their public-interest missions."*[6] The message on who owns the data is hardly coded. The clear implication is that data are of the people and for the people, they're *"the people's data"*.

Although the existing legislation carefully sidesteps the controversial issue of ownership, individuals have a big say on their data and general guidelines for access rights and limitations of use have already been developed, including the principle of proportionality. These are used in the early forays into accessing business data by several countries, as early responses to privacy concerns. A more comprehensive framework is sure to follow. Then, all that will be left would be fostering a 'market' and carving 'ethically-aware paths' that will enable 'scalable, responsible and sustainable' sharing of business data by the public sector. This will lead to a *"sensible, inclusive and participatory data culture through a set of viable, practicable and scalable welfare-enhancing solutions"*.[7] In fact, the European Commission is practically in the final stages of contemplating legislation that will enable governments to access such data, much like it happened earlier with administrative data. Just in case of doubt: *"While, through the open data policy the public sector is required to make its data available, there is no similar policy through which the public sector can access or reuse private-sector data"*.[8] The 'people's data' won't be denied. Incidentally, the EU doesn't favour overt payment by government to businesses for such data. Considering its pioneering General Data Protection Regulations (GDPR) of 2018,[9] with its huge impact on how businesses worldwide treat data belonging to individuals, it won't be accused for not caring about privacy.

The World Bank dedicated its 2021 flagship publication on exactly such matters too. *Better Data for Better Lives* is a clarion call to action for governments, businesses, and all parts of societies to create enabling frameworks for the productive use and reuse of 'public' and 'private' data. *"The innovations resulting from the creative new uses of data could prove to be one of the most life-changing events of this era for everyone"* states the report before it proceeds to make a full-throttled connection between data and socio-economic development.[10] Use of data, from primary to recombined or repurposed, is counted upon to improve the economic and social outcomes of the poor through three institutional pathways: use by governments and international organizations in support of evidence-based policymaking and improved service delivery; use by civil society to monitor government policies, and by individuals to monitor public and commercial services; and use by the private sector in a way that it'll fuel growth for them and, by extension, the whole economy. Governance through an *integrated national data system* tops such aspirational visions, coupled with nothing less than a *social contract* among all players that will enable the realization of the benefits while safeguarding privacy.

### Playing in the Sandbox

To take matters further, we can examine explicitly some of the implications of the new environment on data holders and data producers, including statistical offices. Referencing the situation of a 'datafied' society on behalf of Eurostat, the statistical office of the EU, Fabio Ricciato, Albrecht Wirthmann, and Martina Hahn argue that *"…the changing characteristics of the new data call for a profound rethinking of the official statistics production model…statistical systems must augment their working models, operational processes, and practices...*"[11] They proceed to state that incremental adaptation is not sufficient in this historical moment and call for an evolutionary leap,[†] that would involve wholesale changes in organization, processes, and practices.

The structural change underway is somewhat reminiscent of the experience in manufacturing, with the separation of design from (out)sourced parts. Seeing the production of statistical outputs in a holistic sense, the producer is responsible for the end product and its overall quality but the production of component parts is outsourced elsewhere. In that setting, end-to-end statistical production involves collaboration among many players. The statistical office doesn't need to acquire the data of a private business, it only needs to pretend it has them and act accordingly. It can produce the specs and/or write the code (the black box), which can then be run by the business. The processed intermediate outputs (white circles in Figure 10.1) would be produced there rather than in-house, and the authors state that "*...where the code runs remains independent from what the code does*".[12] Such a model has several implications. Pushing computation out implies loss of control, with the need to manage increased dependencies and vulnerability. However, it doesn't lead to loss of control over methodologies or end outputs. The Apple brand name is still on top of generic components used in the production of its products. A statistical office can still stamp its hefty imprimatur on its outputs.

Similar arrangements can be extendable to other public or private data producers, including researchers. Comparable experiences already exist from past approaches to microdata access. Researchers at times ran regressions without having access to the real data. At times, their code was even run for them by the source, returning results vetted for confidentiality. While this is not mentioned as a best practice in the conduct of research, such approaches are also more conducive to the protection of confidentiality since data don't change hands and only the source 'sees' non-anonymized data.

In the spirit of the microdata discussion in Chapter 8, processed outputs must contain sufficient granularity to explain movements in statistical series. Regardless of where the processing and aggregation of granular data take place, those black boxes require "*thicker methodologies and longer analytic workflows*", which call for the making of several decisions. This "*amplifies issues of methodological sensitivity and subjectivity, and reinforces the need to exercise a critical view on the methods and algorithms (and code) that are eventually set in place and applied to the data*".[13] As we've discussed, alternative data uses may need different algorithms. Over time, alternatives may even be desirable and dictated by the dynamic nature of the source data. Even well-established registers will be due for an overhaul once in a while, a redesign.

Even so, the structure of those intermediate data needs to be as much *input agnostic* and *output agnostic* as possible, that is, independent from the detailed

---

[†] They see the practical implementation as moving from 'official' statistics to Trusted Smart Statistics.

characteristics of the input data that will vary across data sources and time, and from their particular application and use. In other words, processed files must be as multipurpose as possible rather than serve a particular need by one user. For this, processing methodologies will need to exploit synergies. Since, unlike traditional methodologies, they'll include specs translated into code they'll become more modular, resembling software. Individual modules can then change or be replaced without affecting the rest of the processing workflow. Ricciato et al. note that *"The identification of statistical methodology with software code may sound weird to those professionals that interpret these items as ontologically distinct matters"*.[14] Such fusion will produce a cross-breed between methodologists and programmers in a brave new data world!

All these raise the obvious questions why would private businesses do any of that, and who's going to pay? Realistically, does any part of such hypothesized arrangements stand any chance to be part of the future of data? I see the answer as a qualified 'yes', depending again on our own actions. First, the emphasis is not on businesses at large but on big data holders of the type we discussed earlier, and which do have actual or potential statistical capabilities. Some can already be considered statistical outfits. They do similar things for themselves anyway. Second, much like Facebook and others who found themselves having an oversized influence in our societies than they initially envisaged, the businesses who have become big data holders also have inadvertently shouldered more societal responsibilities. It so happens that we move towards corporate cultures where this matters more than it used to. Early experiences of such collaborations bear this out. All that may well be stimulated further by softer measures, raising more awareness of the new issues and the logic behind them. Harder measures, including legislative, are not out of the question either as we saw. Third, with a bit of imagination, mandatory business surveys can be thought of as the black boxes we discussed, the equivalent of running code. Businesses are effectively asked to do something similar in survey questionnaires – process and aggregate across their data. We saw earlier that what constitutes a survey can be pliable, and this could be a survey in the form of the output after running the code! Heavily regulated financial institutions, in particular, fill volumes with required reports for regulators, statistical offices, and central banks very much along similar lines for decades. The eventual routine of running code may actually be an incentive. For the same reasons, big businesses were amenable and willing to try such ideas even before the Internet, under Electronic Data Interchange where they'd give access to parts of their computerized systems in exchange for not being bothered. Similar approaches are of course the norm inside statistical organizations for the use of data from different areas. Fourth, they may do that for commercial reasons too, especially under possible frameworks enabling data markets.

As for compensation, several options can be considered – unlike the case of mandatory surveys. The most unlikely option would be a market price for the data, which doesn't quite exist in our underdeveloped data markets. What is certainly in the offing is compensation for the full recovery of the costs incurred, for human and computing resources. For existing statistical production, such funds can even come from survey savings. Depending on the extent and the exact nature of the required collaborations, other schemes can be contemplated over time, including tax incentives. Compensation can also be in-kind, as in early experimentation with statistical offices, through statistical or analytical products useful to the data-supplying business. For instance, under COVID-19 the French statistical office obtained free data from online accommodation platforms and in exchange it provided them with the aggregated data. As each of them only knew their own sales, this allowed them to estimate their market shares.

## A Trusted Third Party

We now have enough to take our discussion a few steps further and address additional questions that hang around in the new world of data. One such question, frequently asked, relates to how data change hands. This is directly linked to where the data reside. Influenced by practices of the past, even consummate data users tend to equate using the data with 'having' the data. We still exchange data files after all. Not only hand-offs have been common, they've been the predominant means of passing data around. From hard copy printouts, to floppy disks, CD-ROMs, or URLs for downloads, data were moving from one hard drive to the next. In the process, multiple copies were generated. As the sharing and use of all the new data is contemplated, the discussions implicitly concentrate on 'access'. After working hours, with unanimous consent, one can fantasize of models where all the world's data are replicated and housed in a central repository, properly curated and maintained by conscientious caretakers who also arrange for orderly use by accredited researchers and other personae gratae. Rather than a daring futuristic thought, this would be one more dystopian view of the benevolent superintelligence type. On the plus side, I'm not aware of any such requests ever made – for raw data. From administrative data down to the new data, requests are for specified data sets and, increasingly, for access. There's no reason today why huge data sets should leave their location, which is increasingly the cloud. Moreover, data reside more in data environments than data sets. Even from registers of the type we described, desired data outputs can be generated on demand rather than 'exist' in them. Still, processed and anonymized data sets capable of supporting a broad variety of uses can be housed centrally – somewhere.

This links to our previous discussion regarding business incentives for the production of processed files. At least a couple of forces at work warrant more explicit discussion. One is the need for data sets complete with the holdings of all players in a subject matter domain, and which are well designed, updated, and maintained over time, striving to achieve star status. It's such data sets that will bring out the power of the data. They'll confer benefits to the businesses involved but they'll be particularly sought out by users, like a precious commodity. Search engines, social media, and others are effectively already there but their holdings are unique. Who will do something similar for banks, retailers, or mobile phone companies? Will it be industry associations, perhaps? Some may, others may not. Even more challenging, to what extent can such businesses venture outside their core competencies and create autonomous statistical outfits, capable of handling both the management of access and the needed user customization of data sets? For instance, it's one thing for a retailer to enter an agreement with a statistical office for a specified data output but it's another story to satisfy similar demands from other parts of government, private businesses, and researchers knocking at the door. Realistically, how much of that work can be expected to be carried out even among those willing and able?

The second issue concerns the huge costs around the many and diverse activities involved, from managing multiple relations, to sorting out methodological and technical matters of data confrontation and reconciliation at the transaction level, to satisfying user specs. The statistical office will have to negotiate with each and every retailer. The next user will have to do the same. Other businesses, who are both data producers and users, as well as public entities with substantial data holdings, will find themselves in the same predicament too. Very soon, dense and web-like bilateral dealings like those in the left part of Figure 10.2 will become unsustainable. All players will have a vested interest in a more coordinated and orderly approach, one that will maximize collective benefits and minimize costs.

**FIGURE 10.2**
A possible trusted third party in the future world of data

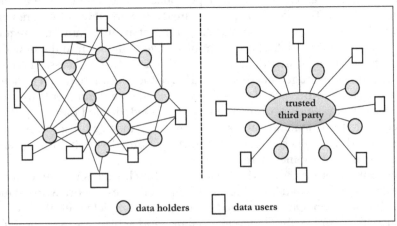

Combined with the need for processed files with wide applicability and our discussion on the location of data, all that points to the desirability of a new entity in that space – a trusted third party. In some countries, it's conceivable that the statistical office may assume some of that role but a more intuitive choice would be a new type of entity situated somewhere between the public and the private sector, in the spirit of public–private partnerships. In addition to being a caretaker of relations and files, the new entity could become a lead player, actively identifying sources, negotiating processes and rules of conduct, establishing standards, curating and housing processed data and metadata, and managing access on behalf of all. Over time, it may even produce data. It would have a key role in the management of a strategic societal resource. This double hub-and-spoke model would look more like a sea urchin, as shown in the right part of Figure 10.2. Ideas related to federated data and other schemes have occasionally been kicked around. The World Bank report emphasizes the importance of 'integrated national statistical systems' involving all players. The EU recommends the creation of national data stewards for the same purpose, as well as for general societal awareness.

With or without such a new authority, the data agenda will move forward in the coming decades. As experiences accumulate and protocols are developed, progress will solidify. New register-like data sources will be created, mapping pathways for newer data to flow seamlessly – as it started to happen in advanced registers today. At more distant times, data linkages will be solidified through automated approaches across all levels, including the raw microdata, in a way that everything talks to everything else. Then, in the event of a crisis, some members of the society will have access to all the knowledge our data can muster.

## More Reflections

This new data world will be very different than the one we knew. We cannot visualize all the changes, and not everything will happen at once. Interim developments, and our own responses along the way, will lead to adjustments, again and again. Here are a few remarks for things to keep an eye on from now on.

*Two-way flows between public and private sources:* Not only business data but government data will be in the spotlight too. Processed data sets will be produced from areas that are not even known yet and which, historically, have not been among the top-tier administrative data. Their holders do not necessarily want to be in the data production, dissemination, and user support business. They will happily delegate to statistical offices or someone else. Expect more two-way data traffic between business and governments.

*Who produces what?* In the new context, with broad data sets processed and available, perhaps housed by a trusted third party, this becomes a very real question. If the CPI can be produced entirely from a variety of centralized and common sources, it's not a far distance before someone other than the statistical office can produce it too – with a proprietary methodology. Ditto for other known outputs. The trust that has been critical in the use of statistics by governments will need additional cultivation, including through the behaviour of governments themselves. Things like that are already happening. The Billion Prices Project was an initiative of MIT and Harvard, collecting data daily from online retailers around the world.[15] Active from 2008 to 2016, it didn't replace the CPI but if we have many more cases of disrepute among statistical offices because of undue political interference the scales can tip and the relative appeal of such alternative statistics can change fast. It'd be confusing to shop at will from a menu of numbers for the same thing – but it may happen. While it's unlikely for governments to lose their appetite for the GDP, even though well-being measures and the like are bandied around for some time now, many new products will appear in an expansive shelf space.

*More accuracy:* Our digital times will be more accurate as we've seen throughout this book. In the past, individuals were effectively asked to perform the function that a black box performs now. How else could they provide estimates of their annual consumption of beef and milk, or their cable bills? Effectively, computation was pushed out to the respondents and the onus was on them to aggregate across multiple transactions, timelines, and family members. We accepted their best-effort answers. More precision is to be expected from data on actual transactions than answers based on the aggregated recall on behalf of household members and across supermarkets or farmers' markets.

*Interim approaches:* In the fusion of the old and the new paradigms, there will be room for several approaches that have not surfaced yet, such as combining traditional sampling with big data. While census-like data may exist in new data sources, there may be instances when the intended data needs may not justify the required processing effort. Nothing prevents the statistical office from designing a survey with a representative sample from a good frame, preparing the questions that need answers and developing the methodology and processing environment to produce the desired output. But then, instead of actually administering the survey, the answers can be populated from the available data sources. For instance, arrangements can be made for the data of a group of individuals from a bank or a utility company. This could combine the best of both worlds as it will not be subject to the criticism of non-representativity, avoid response burden, and lead to better quality until more progress is done and trust is built.

*Collateral damage:* On a high scale, collectively we'll be able to do so much more and better in the future but some things will suffer. What would have been a quick and straightforward answer from a household survey question in the recent past could now become a convoluted and onerous affair if not covered by already-processed data through existing code. The flexibility of discretionary human interventions to 'cut corners' will also be more limited. Getting to some seemingly easy answers will be more roundabout and frustrating than it used to be. This has started to show, perhaps because some of the early steps in the transition have already altered production processes, removing pieces from existing structures without full replacements yet.

*Data ethics:* Rather than relying on good faith alone, roles and responsibilities among the parties will need to be formally defined and managed. Enablers to make the right things happen will need balancing by soft and hard safeguards to protect confidentiality, privacy, and data security. A trusted third party could play a pivotal role here too. For example, pushing computations out could be supervised to ensure the legitimate interests of all stakeholders, including the preservation of the methodological quality of the statistical office, the respect of the business interests of the private data holders, and the protection of citizens' civil rights. Treading thoughtfully in the earlier phases will strengthen the foundations for acceleration later.

For a long time there will be experimentation, with successes and failures. Mistakes will happen and breaches will take place. Everything that can be encountered will be encountered. A sure prediction for the future has to be the element of surprise. Things that look likely may not happen and things that look unlikely may happen.

## Revisiting Data Confidentiality and Privacy

The protection of confidentiality and privacy have been entrenched in statistical practices over the decades. In the spirit of erring on the side of caution, such practices frequently extended beyond the required due diligence. Even elements of occasional paranoia had set in, finding cover behind 'better safe than sorry' attitudes, to the chagrin of users who despise suppressed data cells in statistical tables.

Legislation governing the collection, processing, and dissemination of data is meant to reassure everyone with explicit and strict confidentiality provisions. Business proclamations do likewise, frequently on ethical grounds. The amount of detail surrounding these issues is vast. Volumes of literature can be consulted on any aspect of their legal context and their methodological or technical details. Our brief discussion here will revolve around two matters: how do existing notions of confidentiality and privacy relate to the explosion of data and the new statistical ecosystem, and raise the need for evolution in our collective thinking on such matters as we move ahead.

With respect to the first, there is no doubt that from now on the risks will be higher. Microdata are distributed widely, including among many businesses. The processing steps required for the production of statistical outputs, with or without the direct participation of businesses, will necessitate more linkages and hand-offs – even if the latter happen in the metaphorical sense. True commitment to confidentiality and privacy doesn't start at the dissemination phase by suppressing sensitive cells. Thoughtful practices need to be embedded from the start and adhered to throughout the processes in Figures 10.1 and 10.2. A first step is the anonymization of the data from the very beginning, replacing real identifiers with fictitious and meaningless ones and then 'locking' the keys. This can also be combined with individual consent in those early stages, as needed. For example, someone can have accounts with two or more banks or multiple cell phone numbers, something that will impact data linkages. Higher data layers can be produced subsequently without identifiers. Anonymization goes a long way towards protecting confidentiality and privacy, particularly in data describing individuals.[‡]

Businesses are not entitled to privacy and, generally, anonymization does not provide sufficient protection against the disclosure of their confidential data. The reason is the structure of the economy, with well-known businesses dominating specific sectors and easily identifiable by key statistics, such as employment or revenue size. Accessing and using business data are subject to more restrictive conditions and take place under controlled environments. In ongoing statistical production, a number of mathematical and statistical techniques have been devised to protect confidentiality, ranging from rules requiring cell values to be composed of a minimum number of firms to the degree of the transformation that actual business data have undergone. Even the perturbation of data among cells in a way that it camouflages the real numbers without significant loss in information value has been tried, something not conducive to the new paradigm as it will sow confusion in modular statistical production that relies on re-combining data. Frequently, even for the publication of some aggregate table cells, the consent of dominant firms has also been sought.

Additional safeguards are in place inside statistical organizations. Regardless of the fact that legally all employees can access confidential data, systems of internal approvals have been restricting access on a need-to-know basis. The application of conservative standards for confidentiality and privacy, while warranted, must still be seen under the light of risk management rather than allowing for illusions of absolute mischief-proof masking. After a point things become intractable and practicality needs to prevail, much like crossing the street rather than staying locked inside. With full knowledge of ever-expanding production of more granular statistical outputs showing data from multiple angles, it's not inconceivable that a brilliant and

---

[‡]   The GDPR uses the term *pseudonymization* to differentiate instances in which the linkage keys exist somewhere and re-identification is possible from cases in which anonymization is accomplished in a way that, in principle, re-identification becomes virtually impossible. In the latter case, the processing and storage of personal data would fall outside the purview of the regulation. In reality, however, the boundaries between the two notions depend crucially on the techniques used – and this represents work in progress.

evil mathematician, with considerable programming skills and nothing else to do in life, can unmask something. Aside from hypothesized sinister intents, with the broadening and the intensity of data activities foreseen in the future, mishaps will be inevitable and are bound to happen. This will give rise to additional mitigation protocols, both generic and domain-specific. As part of its '$1 million contest' that Netflix launched in 2006 to improve its recommendation system, it disclosed insufficiently anonymized information for almost half a million subscribers. Cross referencing with other sources made it possible to pigeonhole and identify individuals, including a lesbian mother. In the Doe vs. Netflix class action lawsuit, launched in 2009, she's described as someone "...*who does not want her sexuality nor interests in gay and lesbian themed films broadcast to the world, seeks anonymity in this action*".[16] A whole industry has been built around such matters. Based on all real experiences so far, though, more dangers lurk in security breaches. Compromised and hacked databases with confidential and personal information are headline news all too often. Whether in financial businesses, manufacturers, hotels, or technology companies themselves, it's security vulnerabilities that pose more real risks.

Our second matter is rooted in the argument that legislative statutes and professional ethics are necessary but not sufficient conditions for the functioning and particularly the evolution of the new data environment. A more supportive cultural reawakening with change in societal attitudes will be needed. What lies ahead is simply a big leap away from what's left behind. It can't be sustained by misconceptions or half-understandings of the past, when such issues were addressed in narrow circles through an ostrich approach. Now they need to be taken out in the open with 'adult' conversations. This may well have started, if we perceive the EU's 2018 GDPR as the beginning rather than the end of such a dialogue.

The GDPR is said to be the most assertive and sweeping reaffirmation of individuals' rights to data privacy and security ever. While meant to protect Europeans, its impacts are felt worldwide. Companies that operate in multiple jurisdictions must abide by the regulation in their European operations. More importantly, it already exerts significant influence as a benchmark standard in other jurisdictions. The regulation confers several rights to individuals, such as the right to be informed, to move data anywhere (portability), and to be forgotten (data erasure). It also requires *unambiguous consent* by individuals for any processing of their data. Companies are obligated to protect personal data *by design and by default*, and implement security measures such as two-way authentication and end-to-end encryption. Hopes are also raised that privacy-enhancing technologies will come to the rescue. Moreover, the regulation requires a *data controller*, who decides why and how personal data will be processed, and a *data processor*, who will process data on behalf of the controller. Social media and many other companies can be both since they are by now in the data business, as we've discussed. Stiff penalties are imposed for non-compliance or misbehaviour.

While this legislation not only strongly affirmed but elevated privacy rights historically, there is definitely an air of uneasiness regarding what steps should be taken and by whom as the new data paradigm takes hold. This is not helped by divergent views that still prevail, partly because of notions and practices not widely understood in the past and partly because of different beliefs for the future. As a student of privacy, I've been exposed to public discussions and engaged in private conversations over the years. At times, I felt that my notion of privacy is just that – mine. This is rather strange since we all read from the same legal texts. I've heard information experts distinguishing carefully between freedom of movement and expectations of passing incognito in public, or reassuring people not to worry about their deeply held secrets. The days of the guy going to a faraway corner store for a VHS tape are gone. If some individuals had their way the old telephone books would have never been published and no one would know anything about anyone – no chance for neighbourhood watch initiatives. Others with nonchalant attitudes would passively acquiesce to many intrusions under some 'nothing-to-hide' self-righteousness. People who feel violated if they're asked to provide identification by the police on the street are perfectly conditioned to produce their driver's licence and proof of car insurance. What's perhaps more consequential is that privacy practitioners find it increasingly difficult to reconcile opinionated perceptions of how sacrosanct privacy is with our actions that give away so much and so freely to social media. While our behaviour there cannot be taken as a conclusive societal verdict, it does contain many votes.

As societies we've made sure that our public spaces are practically under surveillance. No sooner than we step outside our homes, we're all monitored. From our street, to the corner store, the shopping mall or the square, all the way inside buses and taxis, cameras are ubiquitous. We may not smile at them but we do count on them to solve crime. Because of such a camera, someone in Canada got a speeding ticket at home from driving a rental car in the Alps. The license plate was traced to the rental company and the remaining records were found in the paperwork there. We've come to count on Google to track our movements. So much has changed, so suddenly, that scrambling to unravel the implications of the new when those of the old weren't that well understood is not surprising. Upfront and honest explanations of the new data context will help remove false dilemmas and avoid unintelligent arguments and inefficient duplication of efforts. Reaching new levels of maturity would require an understanding shared much more widely.

We don't question that our bank has our financial transactions or that the statistical office has our survey responses. However, if we're asked to consent to the sharing of our financial transactions with a 'trusted third party' we'll likely object. But on what grounds? What if the question was *"You've been selected for a mandatory 60-page questionnaire where you need to record every purchase you made over the last 6 months in minute detail. Of course…if you prefer to consent that we access only your transaction data from your financial institution that wouldn't be necessary."* The point is not one of intimidation, of course, but that overt and deeper discussions of what's involved are needed when data are seen as a common shareable resource. A particular practical implication would be to identify parts of statistical registers that can be made public.

Historically, information of the type found in business directories could not be freed from a business register for wider use. Overly restrictive applications of confidentiality masked anything that could identify a business, including information not provided by the business or in the public domain. Identifying a new hotel in a certain area has been a no-no, regardless of the fact that its owners pay to advertise it. Open data along such lines are already being released, though. The financials of publicly traded companies are all disclosed without any loss in competitive advantage. What other 'strategic' information is there that business may only share with an authority, and for how long would that be confidential? Is there a satisfactory cut-off of a few years, analogous to the logic behind the release of the census microdata? And what about the data of businesses that have long gone bankrupt? All these point to some necessary 'dusting' of past legislation.

There's plenty to discuss but from now on the conversation should not only extend to more stakeholders but to the society at large. At the end, it's our data. Separating emotional reactions from rational actions will help balance the protection of privacy and confidentiality with societal goals. Seeing data as a strategic resource is a message that needs to diffuse throughout our societies and gradually seep in. Only then the responsibility will be distributed and the resource will be truly managed collectively.

## The End of Theory

Research and analytical activities to illuminate policy issues, solve business problems, or satisfy human curiosity have been overwhelmingly driven by the *scientific method*. Defining and refining research questions have been imprinted in the psyche of scientists and researchers. Such questions, in turn, are derived from underlying theories or linked to some framework or model, as explained in earlier chapters. Data that can help answer these questions and test the hypotheses formulated are then gathered, with their collection frequently customized for the purpose at hand. Of course, research can start with the data and ask questions after, as we also saw earlier. The new data expand the research possibilities, something that can be interpreted as leading the theory. Either way, the nature of research is such that we'll never run out of questions! This is especially true in socio-economic research, where the portability of findings depends crucially on the context of the time and the place of reference. The arrival of massive amounts of new data in the scene means that much more data-driven research should be expected. Simply 'looking' at the data can reveal patterns.

This is what's behind *the end of theory*. A brief 2008 article by Chris Andersen stirred quite a bit of controversy, reaction to which still continues.[17] He argued that the traditional reliance on models, some of which might have been wrong, is no longer needed since *"the data deluge makes the scientific method obsolete"*. Why bother with all those theoretical dimensions that confuse things to no end? The sheer volume of data makes them *dimensionally agnostic*. Since we can no longer really visualize them, we may as well view them purely mathematically and leave any context for later. No need for semantic or causal analysis, just let the data do the talking. *"This is a world where massive amounts of data and applied mathematics replace every other tool that might be brought to bear. Out with every theory of human behavior, from linguistics to sociology. Forget taxonomy, ontology, and psychology. Who knows why people do what they do? The point is they do it, and we can track and measure it with unprecedented fidelity. With enough data, the numbers speak for themselves."* Research can advance without models.

If you ever wondered about the fan base of models, you'd get the answer from the volume of responses that rushed to their defence. We must postulate rational connections within the data. Data without a model are just noise. Surely, if you torture data long enough, they'll confess to something. But what would the meaning of that something be? We've seen many random patterns and spooky correlations that amount to total nonsense. Even worse, false correlations are an increasing function of data size complicating further our real intent to find causality. As finding causality is definitely beyond the reach of the data-only approach, critics point out that theoretical reasoning is the only antidote.

And the critique continues. Brute data size doesn't have an inside lane to the truth. No data are big enough to make up for inaccuracies they may contain. Moreover, in a realistic context characterized by finite capacity, too much data can be as bad as no data. In complex systems, including the socio-economy, this is brought about by *non-linear saturation* where, rather than adding information, more data can lead to information loss (Sauro Succi and Peter Coveney).[18] This happens because data start to contradict themselves, the bad crowds out and even cannibalizes the good, and our trust in them erodes. Still, believers in Google clicks and other big data not only find solace in synergies with algorithms but go on the offensive. The tired old argument *"why teach long division when calculators can execute it flawlessly"* has morphed to another level: why learn anything when you can look it up?

Weighing inductive versus deductive reasoning isn't new. With upfront apologies to philosophers for any misunderstanding, where do Aristotelian syllogisms that underpin deductive reasoning come from? Don't they need fodder from observations and data? Deductive reasoning has had many successes in science. Yet, epistemologically, the fate of theoretical abstractions relied on the primacy of experiments. Quite recently, theory 'imagined' the Higgs boson and vast sums of money went after experimental data that would prove its existence. Perhaps to a larger extent than in hard science, socio-economic theories and data reinforce each other – most times, anyway. The absence of standards with universal applicability keeps them closer.

Consider the example of the income and substitution effects. When wages go up, the opportunity cost of leisure is raised. Other things equal, this will prompt some people to work more. Others, though, satisfied with their income, will choose to work less. The forces are undeniably there but the theory will not quantify the answer. It'll defer to empirical data for which one prevails, when, and where. In that sense, the theory points where to look for the truth, and the data found there become the truth. However, they represent local and tentative equilibrium points, changing among places and over time. More data will just add higher resolution to the equilibrium points but they won't explain much else. Theory must be able to accommodate similar data elsewhere and at different times, as well as describe the pathways to them. There are times that data alone will suffice, they are the answer. Knowing the size of a crowd in a venue between noon and 3 pm may be useful for something. Asking why the size is what it is, and what intervention would be needed to make it larger or smaller, are different matters that need more than data.

While the data-driven approach won't signal the end of theory or the end of learning, it has already opened up additional research choices. There's no doubt that these will be exercised much more. Equally certainly, humans will never give up the quest to find causal relationships between forces and phenomena that pique our interest. Even for that reason alone, theory won't die.

I'll add one more reason why theory is nowhere near its deathbed, and why the scientific method and data-driven research will coexist. Inertia. In a 2006 paper, *"Our Lives in Digital Times"*,[19] I looked at issues of social alienation, community relationships, why we feel busier than ever despite so many time-saving ICTs and other devices, our spending patterns, and the like. But the paper attracted more attention for a section on 'deaths' that never happened. There had been many such premature pronouncements before the end of theory, including the death of time, of place, and even of history. There was also plenty of talk proclaiming the death of paper and the arrival of the paperless office, the death of the post office because of email, the death of business travel because of videoconferencing, and the death of retail bricks because of e-commerce clicks – something with huge potential implications on employment and real estate. All the implied substitutions behind such phenomena existed, were growing, and were picked up by the data. However, the data detected none of the deaths at the time. Those about to be mourned were all doing quite well. Fifteen years on, the verdict is more-or-less the same. This is not to say that there hasn't been profound change, then and now. The composition of mail has experienced a drastic shift, the logistics of retail are a whole lot different, videoconferencing has skyrocketed. The COVID-19 pandemic specifically has been a game changer. However, in such transitions there are long periods when more of everything coexists. More e-commerce and more in-store retail, more videoconferencing and more travel for conferencing. Even in the era of speed, and despite forces pointing linearly to an end, institutions find a way to persist. There is a huge and always underestimated inertia in our socio-economic systems. I'm fully confident that model-based and data-driven approaches to research will coexist for a long, long time.

## Data by Machines, for Machines

The lion's share of the world's data will be generated by machines. Having outgrown a long list of short-lived storage media, data will reside mostly on *the cloud*. Despite what images may come to mind, this cloud is not in the sky. It's really an environment of an enormous number of distributed and networked servers, collectively mine-melding the brain power we bestowed on them.

The processing of data involves communications in the language of the machines – code. Increasingly, so will be the rest of the production stages leading to data outputs. The human touch will be more visible in less-codified analytical activities. There will still be room for 'our' skills, where we think we have an advantage – despite arguments for being prone to confirmation biases and such. Our interventions will rely on ideas, experiences, and tacit qualities, such as intuition, inspiration, and perhaps the occasional instinct. These will have lasting power but, ironically, they'll enable further codification as they'll help sort out data issues that the machines might have missed. Slowly but predictably, real statistical activities will feel pretty much like someone with some VR gear and a joystick in a basement shoots thousands of kilometres away – for real. We're already at the stage of producing almost-automated releases for *elevator economics*, straight from structured databases. Although never the pinnacle of analytical activities, this speaks to the loss of intimacy between someone and some data on a spreadsheet! As such developments intensify, they'll create *resistance movements* both among old-data nostalgics and genuine rebels. They'll also certainly spur the creation of specialized *data boutiques*, for those who know better, have more discriminating tastes, and can afford to pay for them. The longevity of all the above will remain an open question. It'll be determined endogenously from the relative speeds of all moving parts as the system evolves.

Compared to the supply side, there will be more human involvement in the use side of data. We'll still design much of the research, drawing from an expanded menu of choices and getting into more and more detail. However, we'll find out that many results will be subjected to cruel interrogation, which will give rise to repetitive, cross-checking research. This will happen partly because of inconsistent findings from data-driven research carried out in parallel by machines through operators, but mostly because the state of the processed data will be in flux and in need of quality improvements. While unearthing gold nuggets will be possible, it'll be a period with many frustrations. For a long time, capacity for critical thinking will be indispensable but the real wild card will be the overall numeracy of people. Efforts on that front will be expended by various stakeholders but gaslighting will be rampant. Familiarity with the statistical equivalent of the long division would be an asset.

The insatiable appetite of AI for data will continue on and will be mainstreamed everywhere. Through the Internet of Things, it will also veer towards areas that historically have had little to do with statistics. All sorts of applications will be tried, and a few will create a buzz in social media. Machines will get involved in traditional data uses too. For decades now, central banks, ministries of finance, commercial banks, and think tanks feed a fair amount of data into econometric models used for simulations and economic forecasts. These will be enhanced further with new data from more sources, as well as become more automated. In combination with technical improvements to push technologies already used by data producers, and a little more code here and there, whatever data pieces they have a sweet tooth for will also be swallowed by machines. The moment when the GDP, the CPI, or big data of the future are out of embargo, the machines will ingest them instantaneously in their modelling programs, before any human blinks. While analysts were always instructed to think of the audience they were writing for, they were not forewarned that the biggest data users may also be machines. Hopefully, as spectators, we can be on the lookout for anomalies.

The next two to three decades will involve experiential learning involving trial-and-error, incremental build-up of alliances and linked data sets, and consolidation of methodological and technical processes. With both production and use of data increasingly in the 'hands' of machines, and our wise interventions, the more distant future will bring about complete integration. Everything in Figure 10.1 will become one big thing. Processed data from any source will be linked seamlessly to any other, registers will be updated automatically in real time, even all raw data will be talking to each other. And, it would work! Discovering a negative correlation between blood thinners and grapefruit would be kids' play. Perhaps before the end of the 21st century, 150 years after the first scientific survey, someone like Star Trek's *Data*, an android machine, will be asking the computer intelligence to which himself is a part of: "*How many beings are on this planet and how is this market growing?*"

The computer will oblige. Since there will be no keyboard, and the code will be the voice command, all this will be heard loudly by the captain – a human, I think.

---

## Fact-Checking Tips

We've come a long way in our data journey. Our fact-checking too has dealt with several matters of data. Now, a few words about ourselves and our own data are in order. We all contribute to the evolution of data in various ways. Part of that may be direct depending on our individual capacities. We may be statistical workers, working for a data producer or a business with data holdings, we may be data users, researchers, coders, or storytellers. The other part is what each of us brings to the data world as a citizen, something that requires introspection. Two millennia ago, the stoic Epictetus remarked that our anxiety stems not from what's happening around us but from our thoughts about what's happening. Putting ourselves inside the evolution of data can be best guided by another well-known piece of advice credited to the same thinker, not to fret over things beyond our control and focus instead on what is within our control. This includes our thoughts.

*Control what you can control:* Individually, your data, mine, and everyone else's are drops in the ocean. Collectively, they're a precious resource. More than co-owners, we're contributors to its ongoing creation. No intelligence, artificial or otherwise, will step in to devise and administer an elaborate system of access with customized permissions to our data. We must speak up and be heard. To do so, we need to know – ideally, articulate – what to say. Here are three steps.

The first step involves thinking only. Safeguarding your true secrets, take the time and fact-check carefully who knows what about you. Include family members, neighbours, friends, colleagues, the world. How much of that is by name, either because you told them or because they can find out? How many childhood friends have your photo and know your birthday? The statistical office has your census data, your financial institution knows your mortgage, your landlord your rent, the airline your itinerary and airfare, online retailers your purchases, Google knows your whereabouts, and social media whatever you feed them. Cameras are ubiquitous, someone somewhere can find you anywhere, and you've seen in movies how detectives can link flight manifestos with home addresses and bank accounts to solve crimes. If you grow uncomfortable, surely you can take some actions – short of wearing a balaclava. You can evade the census, pay cash (?), scramble some signals, lie in an application form, turn off some cookies, perhaps read all the pages of a fine print. If you go through the suggested fact-check meticulously, though, I suspect that you'll see none of that as much help. Chances are that in your society and way of life, your time and place, most parameters regarding your privacy are beyond your immediate control. This doesn't mean that you should smile at a drone hovering outside your bedroom window but that you need to think critically of what you give for what you get, every day. Thinking about your data, seriously and pragmatically, is not a trivial exercise. It'll place you in the middle of happenings, you'll grow an appreciation for tough calls that you may or may not be able to influence and, ultimately, you'll become part of the way forward. Fact-checking your own situation is a *reality check* without which no subsequent step can be meaningful or productive.

The second step involves critical thinking of a different kind. What data in your list you'd take back if you could? What would you have given more of if you had to? Which of your data out there you wouldn't mind sharing? What is it that your bank knows that you don't mind sharing with the tax office? What is there in your call logs that you'd hide from the statistical office? How would your life be impacted if someone dug in your garbage and found out what you bought? What if all such data were anonymized and shared widely for the collective good? Would it make a difference if they were used by graduate students for term papers or if they were carefully guarded by a 'trusted third party' in a secure safe, with the keys that would identify you in another, to be used only in exceptional emergencies, perhaps to save lives?

As you go about that, the third step is to discuss as much of your thinking as possible and as widely as possible. How do others think? Role-playing would be helpful here. What levers influence people's tolerance for some seriously expanded future equivalent of white-page data? Feel free to bring in arguments regarding payment for your data or anything else you've been exposed to. Such societal brainstorming will advance the public discourse and will make a twofold contribution to future developments. At a high level, it will become one of the determining forces of change, and it's guaranteed to affect the speed of short-term developments. At a more practical level, gains will come from the fact that you'll give an informed and thoughtful answer next time you're asked for consent. As in an election, your vote may be critical. If you spoil it, there will still be an outcome determined by the votes of others – you'll have to live with the consequences.

---

## Notes

1 OECD, "The Future of Business, Survey," 2020, https://search.oecd.org/indus try/business-stats/the-future-of-business-survey.htm and Meta, "Data for Good, Small Business Surveys," https://dataforgood.facebook.com/dfg/tools/future-of-business-survey

2  OECD, "The Future of Business Survey," www.oecd.org/sdd/business-stats/the-future-of-business-survey.htm

3  Microsoft, "Microsoft Releases 12 million Canadian building footprints as Open Data," *Microsoft Bing Blogs*, March 13, 2019, https://blogs.bing.com/maps/2019-03/microsoft-releases-12-million-canadian-building-footprints-as-open-data.

4  Statistics Canada, "Open Database of Buildings, version 2.0," *The Daily*, www150.statcan.gc.ca/n1/daily-quotidien/190301/dq190301g-eng.htm

5  European Commission, Directorate-General for Communications Networks, Content and Technology, "Towards a European strategy on business-to-government data sharing for the public interest," final report prepared by the "High-Level Expert Group on Business-to-Government Data Sharing, Publications Office," 2021, p. 3, https://data.europa.eu/doi/10.2759/731415

6  Ibid., p. 7. [See also European Commission, "Guidance on sharing private sector data in the European data economy," *Shaping Europe's digital future*, April 2018, https://digital-strategy.ec.europa.eu/en/news/staff-working-document-guidance-sharing-private-sector-data-european-data-economy]

7  Ibid., p. 5.

8  Ibid., p. 14.

9  European Union, "What is GDPR, the EU's new data protection law?", *GDPR.eu*, 2021, https://gdpr.eu/what-is-gdpr/

10  World Bank, *World Development Report 2021: Data for Better Lives* (Washington, DC: World Bank. © World Bank.), 2021, https://openknowledge.worldbank.org/handle/10986/35218, *License: CC BY 3.0 IGO*.

11  Fabio Ricciato, Albrecht Wirthmann and Martina Hahn, "Trusted Smart Statistics: How New Data Will Change Official Statistics," *Cambridge University Press, Data & Policy*, 2 (2020), pp. e7, 15, https://doi.org/10.1017/dap.2020.7

    [See also related paper: Fabio Ricciato, Albrecht Wirthmann, Konstantinos Giannakouris, Fernando Reis and Michail Skaliotis "Trusted Smart Statistics: Motivations and Principles," *Statistical Journal of the IAOS*, 35 (2019), pp. 1–15, 10.3233/SJI-190584. Accessed at ResearchGate, www.researchgate.net/publication/337549354_Trusted_smart_statistics_Motivations_and_principles1]

12  Ibid., p. 9.

13  Ibid., p. 4.

14  Ibid., p. 9.

15  The Billion Prices Project, "About us: 'Big Data' for Macro and International Economics," www.thebillionprices project.com/

16  Wired, "Doe v. Netflix class action lawsuit," www.wired.com/images_blogs/threatlevel/2009/12/doe-v-netflix.pdf

17  Chris Anderson, "The End of Theory: The Data Deluge Makes the Scientific Method Obsolete," *Wired, Science*, June 23, 2008, http://archive.wired.com/science/discoveries/magazine/16-07/pb_theory/

18  Sauro Succi and Peter V. Coveney, "Big Data: The End of the Scientific Method?", *The Royal Society, Philosophical Transactions of the Royal Society A* 377 (2018), 20180145, http://dx.doi.org/10.1098/rsta.2018.0145

19  George Sciadas, "Our Lives in Digital Times," *Statistics Canada, Connectedness Series* (Catalogue no. 56F0004M), 2006.

# Index

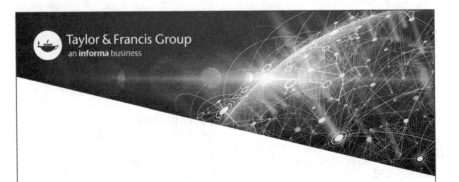

Printed in the United States
by Baker & Taylor Publisher Services

Printed in the United States
by Baker & Taylor Publisher Services